IMPLEMENTING TOTAL SAFETY MANAGEMENT

Safety, Health, and Competitiveness in the Global Marketplace

DAVID L. GOETSCH

Prentice Hall
Upper Saddle River, New Jersey
Columbus, Ohio

Library of Congress Cataloging-in-Publication Data

Goetsch. David L.

    Implementing total safety management: safety, health, and competitiveness in the global marketplace/ David L. Goetsch.

      p.   cm.

    Includes bibliographical references and index.

    ISBN 0-13-243486-5

    1. Industrial safety—Management.   I. Title

T55.G585   1998

658.4'08—dc21

                                                                         97-3804

                                                                         CIP

Editor: Stephen Helba

Production Editor: Christine M. Harrington

Design Coordinator: Karrie M. Converse

Text Designer: Kip Shaw

Cover Designer: Brian Deep

Marketing Manager: Debbie Yarnell

Production Manager: Patricia A. Tonneman

Electronic Text Management: Marilyn Wilson Phelps, Matthew Williams, Karen L. Bretz, Tracey Ward

Illustrations: Christine Marrone

This book was set in Clearface by Prentice Hall and was printed and bound by R.R. Donnelley & Sons, Company. The cover was printed by Phoenix Color Corp.

 © 1998 by Prentice-Hall, Inc.

Simon & Schuster/A Viacom Company

Upper Saddle River, New Jersey 07458

Printed in the United States of America

10 9 8 7 6 5 4 3 2 1

ISBN: 0-13-243486-5

Prentice-Hall International (UK) Limited, *London*

Prentice-Hall of Australia Pty. Limited, *Sydney*

Prentice-Hall of Canada, Inc., *Toronto*

Prentice-Hall Hispanoamericana, S. A., *Mexico*

Prentice-Hall of India Private Limited, *New Delhi*

Prentice-Hall of Japan, Inc., *Tokyo*

Simon & Schuster Asia Pte. Ltd., *Singapore*

Editora Prentice-Hall do Brasil, Ltda., *Rio de Janeiro*

PREFACE

OVERVIEW

Total Quality Management (TQM) has revolutionized the way organizations in the United States do business. As a result, the United States is regaining lost market share in such critical product sectors as automobiles, consumer electronics, and computers. Total Safety Management, or TSM, can play a similarly critical role in helping companies gain and sustain the competitive advantage of a safe and healthy work environment. Just as TQM involves all of an organization in the continual improvement of quality, TSM involves all of an organization in the continual improvement of the work environment. TSM transforms occupational safety and health from a conformance-oriented issue into a competitiveness-oriented issue. The philosophy underlying TSM is that a safe and healthy work environment can give an organization a sustainable competitive advantage. This is because employees are better able to achieve consistent peak performance—one of the keys to competitiveness—if they work in a safe and healthy environment. TSM was first introduced by the author in his book, *Occupational Safety and Health*, 2nd ed. (Prentice Hall, 1996).

Implementing TSM is a fifteen-step process divided into three broad categories of activities: preparation, planning, and execution. It involves employees at all levels of an organization from executives to line workers. With this concept, an organization's health and safety program is no longer viewed as a necessary evil or an unwanted appendage tolerated solely for the sake of compliance. Rather, with TSM, the quality of the work environment becomes a key element in an organization's formula for competitiveness, and safety becomes part of everybody's responsibility. This allows safety and health professionals to become catalysts, facilitators, and coaches. With these professionals providing the spark, the entire organization gets involved in continually improving the work environment—a process that goes on forever.

INTENDED AUDIENCE

Implementing Total Safety Management is intended for both the academic and business/industry markets. In the academic market, colleges, universities, and technical schools that offer programs or courses in occupational safety and health will find the book useful. It can be used as a supplement to a traditional safety and health text (such as my *Occupational Safety and Health,* 2nd ed., Prentice Hall, 1996), or as a stand-alone book in a separate course on health and safety management. In the business/industry market, organizations struggling to meet the challenge of global competition will find the book useful in making TSM part of the solution.

SPECIAL FEATURES

Implementing Total Safety Management is presented in a practical, how-to format that can be used as an annotated model for implementing TSM in any organization. To maximize the instructional value of the book, the following special features are included:

- *TSM Tips.* Each chapter contains three or more TSM Tips. Each tip consists of boxed information relating to some aspect of the chapter content or to workplace safety and health in general.
- *TSM Case Studies.* Each chapter contains a "What Would You Do?" case study that presents a simulated problem facing an individual involved in implementing TSM. Each respective case relates directly to the content of the chapter in question. Readers are asked to review the facts surrounding the problem and decide what they would do in such a situation. Through this simulation and role playing, readers can gain valuable practical experience.
- *Serialized "In Action" Case Study.* Each chapter contains one installment of a serialized case study that illustrates one of the steps in the TSM implementation process *in action*. The serialized case study is the story of how one safety and health professional guides his company through a complete TSM implementation.

ABOUT THE AUTHOR

David L. Goetsch is Provost of the joint campus of the University of West Florida and Okaloosa-Walton Community College in Fort Walton Beach, Florida. Dr. Goetsch is also professor of Safety Management and Quality Management. In addition, he administers the state of Florida's Center for Manufacturing Competitiveness that is located on this campus. Dr. Goetsch is also President of The Management Institute, a private consulting firm dedicated to the continual improvement of organizational performance. Dr. Goetsch is co-founder of The Quality Institute, a partnership of the University of West Florida, Okaloosa-Walton Community College, and the Okaloosa Economic Development Council. He currently serves on the executive board of the Institute.

ACKNOWLEDGMENTS

The author acknowledges the invaluable assistance of the following people in developing this book: Faye Crawford for word processing of the manuscript; Ben Shriver for his helpful editing; and Steve Helba for his vision.

CONTENTS

Overview of TSM

The concept of Total Safety Management (TSM) grew out of a need to transform safety and health management from a strict compliance orientation to a performance orientation in which compliance is an important issue, but not the only issue. The concept was introduced in my book, *Occupational Safety and Health* (2nd ed., 1996, Prentice Hall). A safe and healthy work environment should do more than keep employers out of trouble with regulatory agencies. Pressure from state and federal agencies in the area of workplace safety fluctuates in accordance with the prevailing political climate. But the need to maximize the performance of employees and organizations is constant. Safety and health should be a key element in an organization's plan for gaining a competitive advantage in the global marketplace. TSM is an approach to safety and health management that is rooted in organizational performance and global competitiveness. Its purpose is to give organizations the sustainable competitive advantage of a safe and healthy work environment.

WHAT IS TSM?

TSM is to safety and health management what TQM (Total Quality Management) is to quality management. TQM has revolutionized the way organizations in the United States do business. As a result, the United States is slowly but steadily reclaiming market share lost to Japan and other countries in such critical market sectors as automobiles, consumer electronics, and computers. TSM can produce similar results for occupational safety and health, thereby making organizations who adopt it better able to gain the competitive advantage needed to compete in the global marketplace.

Just as TQM involves the total organization in continually improving quality, TSM involves the total organization in establishing and maintaining a work environment that is both safe and conducive to quality and productivity. Both concepts are rooted firmly in the need to compete globally.

TSM Defined

The origin of TSM can be traced back to the globalization of the marketplace that began after World War II, but really took hold in the 1970s. The need for TSM was created by the need for organizations to be globally competitive. Consequently, TSM is defined as follows:

> *Total safety management is a performance-oriented approach to safety and health management that gives organizations a sustainable competitive advantage in the global marketplace by establishing a safe and healthy work environment that is conducive to consistent peak performance, and that is improved continually forever.*

This definition contains several key elements that must be understood if one is to fully understand TSM. These element are as follows:

■ *Sustainable competitive advantage.* Every organization that competes at any level, but especially those that compete at the global level, must have competitive advantages. These are capabilities or characteristics that allow them to outperform the competition. For example, if the organization in question is a baseball team, it might have such competitive advantages as an excellent pitching staff, several speedy baserunners, two or three power hitters, and/or outstanding fielders in key positions. These advantages, if exploited wisely, will help make the baseball team a winner. If these advantages can be sustained over time, they will help make the team a *consistent* winner.

This same concept applies to organizations that compete in the global marketplace. In order to survive and prosper, they need as many competitive advantages as possible. Traditionally, competitive advantages have been sought in the key areas of quality, productivity, service, and distribution. However, peak-performing organizations have learned that a safe and healthy work environment is essential to gaining competitive advantages in these critical areas. In fact, a safe and healthy work environment is itself a competitive advantage. In today's competitive marketplace, high-performance employers are adding one more critical area to the list of those in which competitive advantages are sought. This new addition to the list is the work environment.

TSM TIP

Cost of Workplace Accidents

The approximate cost of accidents that occur on or off the job each year in the United States is $150 billion.

Peak performance organizations are learning that a safe and healthy work environment gives them a doubly effective competitive advantage. First, it ensures that employees work in an environment that allows them to focus all of their attention, energy, and creativity on continually improving performance. Second, it prevents an organization's limited resources from being drained off by the non–value-added costs associated with accidents and injuries.

■ *Peak performance.* The primary driver behind TSM is organizational, team, and individual performance. An organization's ability to survive and prosper in the global marketplace is determined largely by the collective performance of individuals and teams. Consistent peak performance by all individuals and teams in an organization is essential to long-term success in the global marketplace. The quality of the work environment is a major determinant of the performance levels that individuals, teams, and organizations are able to achieve. A better work environment promotes better performance.

■ *Continual improvement forever.* People work in an environment, and the quality of that environment affects the quality of their work. The work environment is a major determinant of the quality of an organization's processes, products, and services. In the age of global competition, quality is an ever-changing phenomenon. Quality that is competitive today may not be tomorrow. Consequently, continual improvement is essential. If quality must be improved continually, it follows that the work environment must also be improved continually.

TRANSLATING TSM INTO ACTION

There are three fundamental components through which the TSM philosophy is translated into action on a daily basis. These three components are the TSM Steering Committee, Improvement Project Teams (IPTs), and the TSM Facilitator.

TSM Steering Committee

The TSM Steering Committee oversees the organization's safety and health program. It is responsible for the formulation of safety and health policies, the approval of internal regulations and work procedures relating to safety and health, the allocation of resources, and approval of recommendations made by the IPTs.

TSM TIP

Value of a Strong Steering Committee

"Executives who have created a Steering Committee that works as a team have said that they would never go back to the traditional way of managing and behaving."[1]

Improvement Project Teams

Improvement Project Teams (IPTs) are ad hoc or temporary teams formed by the TSM Steering Committee for the purpose of pursuing specific improvement projects relating to the work environment. Through the ongoing process of hazard identification, in which all employees in a TSM organization participate, potential safety and health problems are identified. When a potential problem is identified, an IPT is formed to analyze the situation and make recommendations for improvements. These recommendations are presented to the Steering Committee by the TSM Facilitator. For example, if an organization's workers' compensation costs have increased sharply, or if the number of accidents in a given department is on the rise, IPTs might be formed to analyze these situations and make recommendations.

Typically, IPTs are cross-functional. This means that all stakeholders in the problem are represented on the IPT that will look into it. A stakeholder is any individual, team, or department that is affected by the problem in question, directly or indirectly. Consequently, IPTs usually have members from several different departments. Hence, the term *cross-functional*.

However, IPTs can also be natural work teams. These are teams made up of people who work together on a daily basis in performing their jobs. How an IPT is constituted is determined by the nature and scope of the problem. If the hazard's effects are confined to a single work environment, the IPT should be a natural work team. If its effects cross into several different work environments, the IPT should be cross-functional.

Once an IPT fulfills the charter it is given by the Steering Committee, it is disbanded. While one IPT is working on an assigned project, others are formed to pursue other projects. This is an ongoing process that never stops because a fundamental aspect of TSM is continual improvement of the work environment forever.

TSM Facilitator

The TSM Facilitator must be a safety and health professional from within the organization. Ideally, he or she should be the organization's chief safety and health officer. The TSM Facilitator serves as the Steering Committee's resident expert on the technical and compliance aspects of safety and health, and is responsible for the overall implementa-

Figure I–1
Tasks and Roles of a Facilitator
in a Group Meeting

- Participate, but don't dominate.
- Promote diverse viewpoints.
- Stay positive—don't be defensive.
- Draw all participants into discussions.
- Promote open, frank interaction.
- Keep the group on track and maintain adherence to the agenda.
- Maintain control.
- Identify and mediate underlying conflicts.
- Discourage overly dominating participants.
- Periodically summarize key points, agreements, and action items.
- Monitor the reactions of participants.

tion and operation of the organization's TSM program. Figure I–1 lists several roles a facilitator must be able to play in group meetings. The better an individual is at playing these roles, the better a facilitator he or she will be.

FUNDAMENTAL ELEMENTS OF TSM

TSM differs in several ways from traditional safety and health management. In order to appreciate these differences, one must understand the fundamental elements of TSM. Fundamentally, TSM is

- Strategically based
- Performance oriented
- Dependent on executive-level commitment
- Teamwork oriented
- Committed to employee empowerment and enlistment
- Based on scientific decision making
- Committed to continual improvement
- Involves comprehensive, ongoing training
- Promotes unity of purpose

Each of these elements is discussed in the following sections.

TSM TIP

Ranking of Industries by Death Rates

On the basis of the number of deaths per 100,000 employees, various industries can be ranked as follows (from the highest death rates to the lowest):

- *Mining/quarrying*
- *Agriculture*
- *Construction*
- *Transportation/publications*
- *Government*
- *Manufacturing*
- *Services*
- *Trades*

Strategic Basis

TSM's being strategically based means that an organization views a safe and healthy workplace as giving it a *competitive advantage*. Consequently, the organization makes maintaining a safe and healthy workplace a part of its strategic plan. Traditional organizations, on the other hand, tend to view safety and health more from a compliance perspective. TSM is strategically based because it is considered part of an organization's overall strategy for competing globally.

In order to compete in the global marketplace, organizations must plan for and establish competitive advantages. This is known as strategic planning. Once the planned advantages have been established, they must be maintained over time and exploited to the fullest possible extent. Strategic planning is a matter of answering the following questions:

- As an organization, what would we like to be? What is our Dream? This is the *vision* component of the strategic plan.

- As an organization, what is our purpose? This is the *mission* component of the strategic plan.

- As an organization, what values are most important to us? This is the *guiding principles* component of the strategic plan. A guiding principle is a high-priority organizational value.

- As an organization, what do we want to accomplish? This is the *broad objectives* component of the strategic plan. Figure I–2 contains an excerpt from the strategic plan for a hypothetical company, Jones Manufacturing.

Jones Manufacturing Company

Vision

Jones Manufacturing Company will be the leading supplier of home-fitness equipment in the United States.

Mission

The purpose of Jones Manufacturing Company is to design and manufacture high quality fitness equipment for resistance training in a home environment.

Guiding Principles

The highest priority corporate values of Jones Manufacturing Company are as follows:

- Customer satisfaction
- Product and service quality
- Cost leadership
- Employee safety and health
- Continual improvement
- Ethical business practices

Figure I–2
Core of the Strategic Plan for Jones Manufacturing

With TSM, a safe and healthy work environment shows up in the strategic plan as a competitive advantage that will be actively pursued and—once established—will be maintained over time. Safety and health concerns appear either in the *guiding principles* section of the strategic plan, the *broad objectives* section, or both. For example, in Figure I–2, employee safety/health is a guiding principle value of Jones Manufacturing Company.

Employee safety/health might also appear in the strategic plan as a broad objective. For example, the following broad objective might appear in an organization's strategic plan:

> *To maintain a safe and healthy work environment that is conducive to consistent peak performance on the part of all employees.*

Performance Orientation

In any organization that is subject to federal and state regulations, compliance is an important issue. Failure to comply with applicable regulations can lead to fines and other non–value-added expenses. Companies in such sectors as manufacturing, processing, construction, transportation, and maritime services must comply with numerous government safety regulations. Figure I–3 contains a partial list of government agencies that write regulations concerning workplace safety. Small wonder, then, that in many traditional companies, compliance has become not just an important issue, but the only issue with regard to safety and health.

In a compliance-driven setting, the safety department has a regulation orientation. Such an orientation can breed a *big-brother-is-watching-you* mentality among employees and managers that can cause them to resent the safety department's efforts. The resentment is magnified when employees and managers are pressed to meet deadlines, and they see safety regulations as slowing them down.

With TSM, the safety department has a performance orientation, and safety/health personnel are viewed as part of the equation for continually improving performance. Compliance is still important, of course. But compliance and productivity are no longer viewed as being mutually exclusive entities. Compliance is viewed as a natural extension of an organization's performance-improvement strategies, and continues to be an important responsibility of the TSM Facilitator.

Executive Commitment

The TSM approach can be implemented fully and successfully only if an organization's executive management team is committed to providing a safe and healthy work environment as a performance-improvement strategy. For this reason, the modern safety and health professional must be able to convincingly articulate the TSM philosophy. If executive commitment does not exist, it will have to be generated. Naturally, responsibility for generating the necessary commitment falls to safety and health professionals.

Teamwork Orientation

In a traditional setting, workplace safety and health tend to be viewed as the responsibility of the safety department. In a TSM setting, although the safety department plays the

Figure I–3
Government Agencies
Concerned with Workplace
Safety

American Public Health Association

Bureau of Labor Statistics

Bureau of National Affairs

Commerce Clearing House

Environmental Protection Agency

National Institute for Standards and Technology (formerly National Bureau of Standards)

National Institute of Occupational Safety and Health

Occupational Safety and Health Administration

Superintendent of Documents, U.S. Government Printing Office

U.S. Consumer Product Safety Commission

key facilitating role, ensuring a safe and healthy workplace is everybody's responsibility. IPTs are formed by the TSM Steering Committee and given charters to improve specific aspects of the work environment. When a team satisfies its charter, it is disbanded. As the need to improve the work environment continues, new teams are chartered, and the process continues forever.

Employee Empowerment and Enlistment

Empowerment means involving employees in ways that give them a real voice. With empowerment, employees are *allowed* to give input concerning workplace issues, problems, and challenges. If the issue is how to make the work environment safer, empowered employees are allowed and encouraged to make suggestions, and their suggestions are given serious consideration. Enlistment takes the concept of empowerment one step further. With enlistment, rather than allowing employees to give input, the organization *expects* them to.

In a TSM setting, employees are viewed as invaluable sources of information, knowledge, insight, and experience that should be tapped continually. Consequently, employee input is not just wanted, it is needed; it is not just sought, it is expected. A criterion that should be included on the performance-appraisal form of modern organizations reads as follows:

> To what extent does this employee provide input that is useful in making continual improvements in the workplace?
>
> Frequently   _____
>
> Sometimes   _____
>
> Seldom   _____
>
> Never   _____

Scientific Decision Making

It is no exaggeration to say that decisions about workplace safety and health can, on occasion, be life-or-death decisions. Whether they are that or not, all such decisions are too important to be based on guesswork. With TSM, scientific decision making is the norm. This means that employees in organizations that adopt TSM must learn to use such decision making tools as Pareto charts, scatter diagrams, histograms, and fishbone diagrams. Such tools can take the guesswork out of decision making and problem solving and, as a result, lead to better decisions and better solutions.

Continual Improvement

With TSM the status quo is never good enough, no matter how good it is. In a competitive marketplace, it must be assumed that the competition is improving continually. Consequently, what qualifies as competitive performance today may not tomorrow. The state of the work environment can always be improved, and it should be. In an organization that adopts TSM, IPTs are always at work and recommending accident prevention strategies that will make the workplace safer and more productive.

Comprehensive, Ongoing Training

In a TSM setting, training is an ongoing part of the job. If all employees are going to be involved in hazard identification and accident prevention, they must be trained in the basic techniques. If employees are going to be members of IPTs, they must be trained in the fundamentals of teamwork. If employees are going to use scientific tools in making recommendations for solving problems, they must be given instructions in the proper use of these tools.

Training should be comprehensive in nature, and ongoing. Figure I–4 contains a partial list of courses, seminars, or workshops that employees in a TSM setting should

Figure I–4
TSM-Related Training
Opportunities

- TSM: What It Is and How To Do It
- Teamwork Fundamentals
- Scientific Decision-Making/Problem-Solving
- Stress and Safety
- Mechanical Hazards and Bodyguarding
- Lifting Hazards
- Falling, Impact, and Acceleration Hazards
- Heat and Temperature Hazards
- Pressure Hazards
- Electrical Hazards
- Fire Hazards
- Toxic Substances
- Explosive Hazards
- Radiation Hazards
- Noise and Vibration Hazards
- Hazard Analysis
- Accident Prevention
- Emergency Preparation/Response
- Facility Assessment/Hazard Identification
- Accident Investigation and Reporting
- Industrial Hygiene
- Automation-Related Hazards
- Bloodborne Pathogens and Related Hazards
- Ergonomics in the Workplace

TSM TIP

Injuries on the Job—Leading Causes

The leading causes of workplace injuries are as follows: overexertion, impact accidents, falls, bodily reaction to a substance, compression, motor vehicle accidents, exposure to radiation or caustic substances, abrasions, and exposure to extreme temperatures.

---

have access to. The specific training opportunities that should be provided are determined by the nature of the organization, its products, processes, and employees.

Unity of Purpose

In a TSM setting, all employees at all levels understand that safety and health are the responsibilities of everyone. The TSM Facilitator and other safety and health professionals are responsible for planning, facilitating, coordinating, advising, and monitoring. But employees are responsible for doing their part to make the workplace safe and healthy. Employees satisfy their responsibilities in this regard by doing the following:

- Setting a positive example of working safely
- Encouraging fellow employees to work safely
- Practicing hazard identification techniques constantly
- Recommending accident prevention strategies
- Serving effectively on IPTs

Gaining unity of purpose begins with the commitment of executive management. An organization's executive management team must commit to safety and health, make it a priority issue in the organization's strategic plan, stress its importance through both words and example, and expect it to be a priority of all employees.

RATIONALE FOR TSM

Does the quality of the work environment affect the quality of an employee's performance? According to a study conducted by the Institute for Corporate Competitiveness (ICC), it does.[2] The ICC surveyed 100 employees from ten different manufacturing firms to determine their perceptions of how various work-related factors affect their performance. Of the fifteen factors listed in the survey instrument, *Quality of the Work Environment* was ranked in the top three by 92 percent of the respondents.

The correlation between work environment and job performance is strong. It is this correlation that forms the basis of the rationale for TSM. Globalization has increased the level of competition in the marketplace exponentially. For many organizations, adjusting to globalization of the marketplace has been like jumping from high school athletic competition to the Olympics.

Organizations that find themselves in these circumstances need every competitive advantage they can muster. They need to do everything possible to continually improve performance. One thing an organization can do—and must do—is provide employees a safe and healthy work environment that is conducive to consistent peak performance.

IMPLEMENTING TSM: THE MODEL

Figure I–5 contains a three-phase, fifteen-step model that can be used for successfully implementing TSM in any organization. The three broad phases of activity are planning

---

Steps in the TSM Implementation Process

Planning and Preparation

1. Gain Executive-Level Commitment
2. Establish the TSM Steering Committee
3. Mold the Steering Committee into a Team
4. Give the Steering Committee Safety and Health Awareness Training
5. Develop the Organization's Safety and Health Vision and Guiding Principles
6. Develop the Organization's Safety and Health Mission and Objectives
7. Communicate and Inform

Identification and Assessment

8. Identify the Organization's Safety and Health Strengths and Weaknesses
9. Identify Safety and Health Advocates and Resisters
10. Benchmark Initial Employee Perceptions Concerning the Work Environment
11. Tailor Implementation to the Organization
12. Identify Specific Improvement Projects

Execution

13. Establish, Train, and Activate Improvement Project Teams
14. Activate the Feedback Loop
15. Establish a TSM Culture

Figure I–5
Model for Implementation of TSM

and preparation, identification and assessment, and execution. Phase One—*planning and preparation*—encompasses Steps 1–7. Each step is completed in order. Pursuing steps out of order can throw the implementation process into disarray. For example, Step 1 is a critical preparation activity. Failure to successfully complete Step 1 can jeopardize all subsequent steps. This is because the organization's attitude toward TSM will be determined primarily by the attitudes of executive managers.

Step 2 is the establishment of the TSM Steering Committee. The ideal TSM Steering Committee is the organization's executive management team, possibly augmented as necessary to ensure that all functional departments are well represented. Steps 3 and 4 are preparation activities that involve training. In Step 3, the TSM Steering Committee undergoes teamwork training and teambuilding activities. In Step 4, the Steering Committee undergoes safety and health awareness training.

Steps 5 and 6 are planning activities in which the Steering Committee develops a mini–strategic plan for safety and health. This plan consists of a vision, guiding principles, mission, and broad objectives, all of which focus solely on safety and health in the workplace. This mini–strategic plan becomes a subset of the organization's overall strategic plan. In Step 7, the plan is communicated to all employees.

Phase Two of the implementation model—*identification and assessment*—encompasses Steps 8–12. The steps in this phase allow the Steering Committee to identify strengths that might work in favor of the implementation, as well as inhibitors that might work against it. Having identified strengths and inhibitors, the Steering Committee can tailor the implementation so as to exploit strengths while minimizing inhibitors.

Phase Three of the implementation model—*execution*—encompasses Steps 13–15. The steps in this phase involve actually assigning teams to specific improvement projects. As soon as a given improvement has been made, that IPT is disbanded and a new team is formed to pursue other improvements. This process repeats itself continually forever. An outgrowth of the repetitive nature of the model is that TSM becomes ingrained in the organization's culture.

SUMMARY

1. TSM is an approach to safety and health management that gives organizations a sustainable competitive advantage in the global marketplace. This is accomplished by involving all employees in establishing, maintaining, and continually improving the work environment such that it is conducive to consistent peak performance.

2. The fundamental elements of TSM are its strategic basis, performance orientation, executive commitment, teamwork orientation, employee empowerment and enlistment, scientific decision making, continual improvement, comprehensive and ongoing training, and unity of purpose.

3. The rationale for TSM can be found in the connection between job performance and the work environment. In order to compete in the global marketplace, organizations need all employees performing at peak levels on a consistent basis. A safe and healthy workplace promotes peak performance.

4. The model for implementing TSM consists of fifteen steps arranged in three phases. These phases are planning and preparation, identification and assessment, and execution.

KEY TERMS AND CONCEPTS

Comprehensive, ongoing training

Continual improvement

Employee empowerment and enlistment

Executive commitment

Peak performance

Performance oriented

Scientific decision making

Strategically based

Sustainable competitive advantage

Teamwork oriented

Unity of purpose

REVIEW QUESTIONS

1. Define the term *Total-Safety Management*.
2. List and briefly describe the fundamental elements of TSM.
3. What is the rationale for TSM?
4. What are the three broad phases that make up the model for implementing TSM?
5. What is the most important step in the TSM implementation model, and why?

ENDNOTES

1. David L. Goetsch and Stanley B. Davis, *Implementing Total Quality* (Englewood Cliffs, N. J.: Prentice Hall, 1995), p. 68.
2. Institute for Corporate Competitiveness, "Employee Perceptions: Impact of Work Factors on Job Performance," Report 95–6, August 1995, p. 2.

Gain Executive Commitment

- TSM as Cultural Change
- Executive Commitment: A Must
- Achieving Executive Commitment
- Maintaining Executive Commitment
- Step 1 in Action

Of all the factors that contribute to successful implementation of TSM, the commitment of executive-level managers is the most important. Without executive-level commitment the implementation is almost sure to fail. This step involves gaining and maintaining the level of commitment necessary for success.

TSM AS CULTURAL CHANGE

According to Peter Scholtes, " . . . culture refers to the everyday work experiences of the mass of employees."[1] Deal and Kennedy call organizational culture " . . . a system of informal rules that spells out how people are to behave most of the time."[2] In other words, an organization's culture consists of its customs, traditions, rites, and rituals.

Organizational culture has several elements as shown in Figure 1–1. If the existing culture of an organization runs counter to the TSM philosophy, it will have to be changed in order for the TSM implementation to work. Effecting cultural change can be a slow, difficult, and sometimes frustrating process. It involves changing three of the individual elements of the organization's culture: its values, its role models, and the complex of its customs, traditions, rites, and rituals. Cultural change also deeply affects how the organization responds to the fourth key element—the marketplace.

Figure 1–1
Makeup of an Organization's
Culture

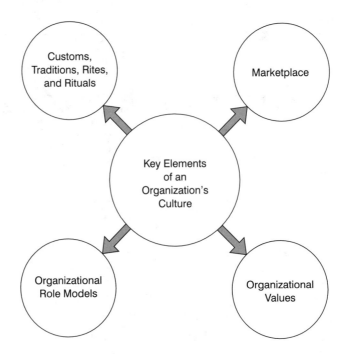

Assessing Organizational Culture

Before attempting to gain executive commitment to TSM, the safety and health manager should assess the current state of the organization's culture from the perspective of safety and health. Is it *TSM-friendly* already, or does it run counter to the TSM philosophy? Figure 1–2 is an assessment checklist that can be used to answer this question.

The checklist is completed by the safety and health manager based on his or her own observations and experiences, input from other employees, and actual evidence that is available in the form of written documents. Items 1–7 of the checklist are the most important criteria from the perspective of cultural change. Analysis of all the items is necessary in order to get a clear picture of the current organizational culture vis-a-vis

---

TSM TIP

What Is Organizational Culture?

"Organizational culture is the set of assumptions, beliefs, values, and norms that is shared among its members. This culture may be conspicuously created by its members, or it may have simply evolved across time. It represents a key element of the work environment in which employees perform their jobs."[3]

Organizational Values

Yes   *No*

_____ _____   1. Does the organization's CEO have a positive attitude toward safety and health issues?

_____ _____   2. Do all members of the organization's executive management team have positive attitudes toward safety and health issues?

_____ _____   3. Do executive managers appear to view the work environment as having an effect on employee performance?

_____ _____   4. Is there a safety-and-health related guiding principle in the organization's strategic plan?

_____ _____   5. Is there at least one broad safety-and-health objective in the organization's strategic plan?

_____ _____   6. Are safe and healthy work practices included as criteria of the performance appraisal process?

_____ _____   7. Do executive managers cooperate in the promotion of a safe and healthy workplace?

Organizational Role Models

_____ _____   8. Do executive managers serve as positive role models of safe and healthy work behavior?

_____ _____   9. Do executive managers consider safety and health issues as applicable when making decisions?

_____ _____   10. Do supervisors/mid-managers consider safety and health issues as applicable when making decisions?

_____ _____   11. Do supervisors point out unsafe behavior and insist that appropriate cautions be taken all of the time?

Customs, Traditions, Rites, and Rituals

_____ _____   12. Is there peer pressure among employees to work in a safe and healthy manner?

_____ _____   13. Are safety and health stressed when new employees are oriented?

_____ _____   14. Are safe work practices recognized and rewarded?

_____ _____   15. Are safety and health discussed by employees during informal gatherings?

Figure 1–2
Cultural Assessment: Safety-Friendly Organization?

safety and health. However, it is Items 1–7 that indicate the extent of executive commitment to safety and health. Without this commitment, the remaining criteria lose much of their relevance.

Planning for Change

The various criteria in Figure 1–2 are *cultural indicators* or individual puzzle pieces that, when put together, form a picture of the organization's culture as it relates to safety and health. These indicators, if properly interpreted, can be used in developing a plan for effecting the necessary cultural change.

In developing such a plan, safety and health managers should be aware that change seldom occurs easily or without conflict. If the necessary change is cultural in nature, the situation is complicated even further. Deal and Kennedy recommend the following strategies for promoting cultural change:[4]

- *Peer group consensus will be the major factor influencing acceptance or rejection of change.* This typically works in favor of safety and health managers who are promoting TSM. Employees are usually supportive of safety and health improvements in the workplace. After all, it is their well-being that is in question.

- *Communicate constantly to build two-way trust.* It is important for employees and managers to understand that TSM is something *we will do* for ourselves rather than something that will be *done to us*. Keep everybody at all levels fully informed. With TSM or any other cultural change, there should be no surprises.

- *Make training part of the change process.* One of the reasons people resist change is fear of personal obsolescence. Providing training can alleviate such fears. One of the reasons plans fail is that people are given implementation responsibilities for which they are unprepared. Training can give employees the knowledge and skills they need in order to do their part in executing an implementation plan.

- *Allow time for change to take hold.* Consider the following situation. For years a busy intersection in a residential neighborhood had no stop signs. Accidents and near-misses were common and becoming worse when concerned residents finally convinced the city council to make the intersection a four-way stop. Thinking the problem solved, relieved residents were surprised to see cars continue to zoom through the intersection as if the stop sign didn't exist. After a couple of weeks, drivers could be seen slamming on their brakes and screeching to a smoking halt instead of running the stop sign. Some even slammed on their brakes after passing through the intersection. Drivers who frequented this particular route took more than a month to finally realize that a change had occurred and respond accordingly. The point here is that change, no matter how positive, takes time because people tend to be set in their ways. If it is true that patience is a virtue, then when trying to promote cultural change, it is doubly so. With a cultural assessment completed, the plan for making any necessary cultural changes can be developed. The first step in that plan must be to gain executive commitment to TSM.

TSM TIP

Categories of Safety and Health Professionals

The modern safety and health professional may be a safety manager, safety engineer, industrial hygienist, environmental engineer, health physicist, occupational physician, occupational nurse, or technologist/technician with a safety/health specialization.

EXECUTIVE COMMITMENT—A MUST

In any organization the tone and direction are set by the chief executive officer (CEO) and the members of his or her executive management team. These individuals decide what will be emphasized, where limited resources will be spent, what behavior will be rewarded and what won't, and how incentives will be used. Mid-managers and supervisors take their cues from executive managers. They, in turn, pass their perceptions along to employees.

The TSM philosophy requires that the *total* organization be involved in continually improving the work environment. TSM cannot be implemented in just one department or by just a few employees. By definition, such an approach would not be *total* safety management, and anything short of total involvement robs the organization of the full benefit of TSM.

Consider just a few of the things that must be done in order for TSM to be implemented:

- Safety and health must be included as high priority concerns in the organization's strategic plan.
- Resources must be allocated to cover implementation costs (e.g., planning, training). Employee performance relative to safety and health must be monitored, evaluated, and rewarded as appropriate.
- Safety and health must be monitored, evaluated, and rewarded as appropriate.
- Employees must be shown that safety is a must no matter how pressed the organization becomes to meet deadlines.

Who but the CEO and executive managers of the organization has the authority to do these things? The answer is *no one*. Consequently, executive commitment is a must.

ACHIEVING EXECUTIVE COMMITMENT

What does it mean to achieve executive commitment? The concept is defined by its three components as shown in Figure 1–3. An executive who is committed to TSM, or

Figure 1–3
Components of Executive
Commitment

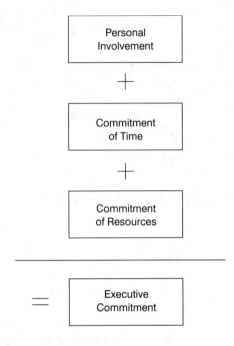

any other concept, for that matter, will be personally involved in its implementation. Figure 1–4 contains a checklist of ways in which an executive-level manager can be personally involved in TSM.

An executive manager who is completely committed to TSM will want to serve on the organization's TSM Steering Committee. The ideal steering committee consists of the organization's executive managers. If the CEO and his or her executive managers do not serve on the TSM Steering Committee, they will have to go to even greater lengths to demonstrate their commitment to safety and health. Otherwise the committee's credibility will suffer.

Figure 1–4
TSM Personal-Involvement
Checklist for Executive-Level
Managers

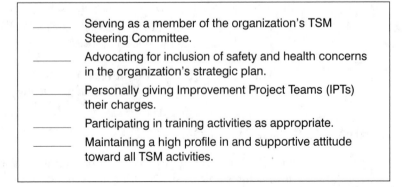

_____ Serving as a member of the organization's TSM Steering Committee.

_____ Advocating for inclusion of safety and health concerns in the organization's strategic plan.

_____ Personally giving Improvement Project Teams (IPTs) their charges.

_____ Participating in training activities as appropriate.

_____ Maintaining a high profile in and supportive attitude toward all TSM activities.

========== TSM CASE STUDY ==========

What Would You Do?

Vinco's CEO, Mary Earnst, is moving the company forward in implementing TSM, but she isn't sure what the process will mean for her executive management team. She knows that each member of the team will have to be personally involved in the implementation, but what does *personally involved* really mean? Earnst wants to know, as do Vinco's other executive managers. If you were Vinco's TSM Facilitator, how would you explain *personal involvement* to Mary Earnst?

The safety and health manager for the organization meets with the executive management team whenever it sits as the TSM Steering Committee, and serves, in such cases, as the committee's facilitator and consultant.

Executive managers committed to TSM will advocate on behalf of safety and health when developing the organization's overall strategic plan. Through their involvement in the process, executives can ensure that safety and health concerns show up in the organization's strategic plan as guiding principles and/or broad objectives, or both.

Each time an Improvement Project Team (IPT) is established to deal with a specific safety or health concern, executive managers can be personally involved by giving the team its charter. Receiving its charter directly from the CEO or another executive manager tells an IPT that the activity in question is important.

Personal participation in the various training activities associated with implementing TSM is doubly beneficial for executive managers. First, executives learn what they need to know in order to play a positive role in operationalizing the TSM philosophy. Second, their participation tells employees that *"this training is important."*

During the implementation of TSM, and after it has been operationalized, committed executives will want to maintain a high profile and a supportive attitude toward TSM. This means conducting/attending meetings, publicly communicating progress, writing editorials or commentaries in company newsletters, publicly rewarding/acknowledging TSM-supportive behavior, and personally visiting work sites regularly. When visiting work sites, executives can question employees about their experiences with TSM, and solicit input concerning improvements that are needed.

The second component of executive commitment to TSM is the commitment of time. Employees know intuitively that executives spend their time doing what they think is important. Consequently, when executives are seen devoting time to TSM activities, the message is conveyed that "TSM is important." Conversely, if executives display an attitude that says, "I don't have time for this," employees will silently respond, "Neither do I."

The final component of executive commitment is the commitment of resources. As important as the other two components are, this one is even more important. Employees know that limited resources are expended in order of priority. Consequently, when employees see executives devoting resources to TSM, they know that TSM is one of the organization's priorities.

All of these components taken together—personal involvement, time, and resources—to the extent that they are present, determine the level of executive commitment. Convincing executives to get personally involved, and to commit both time and resources is the most important challenge the safety and health manager faces in attempting to implement TSM.

Evolution of Executive Commitment

Executive commitment to TSM—with rare exceptions—will take time to achieve. Safety and health managers should expect to confront the evolutionary steps shown in Figure 1–5. Executive managers may be skeptical when they first hear about TSM. This is a normal human response to change, and TSM means change. A persuasive argument, persistent but patiently repeated often enough, may move executive managers to the next level: tentative interest. If so, the safety and health manager will probably be asked to make a presentation to the executive management team. If the presentation goes well, TSM will probably be provisionally accepted.

At this point it's a good idea to undertake a pilot project to demonstrate how TSM can make a difference. For example, an IPT might be formed to confront a specific safety/health problem. If the pilot project goes well, executive managers will probably move to the buy-in stage. Once TSM has been implemented widely and positive results have been demonstrated, executives will probably move to the commitment phase.

Four-Step Process for Gaining Executive Commitment

In order to gain executive commitment, the safety and health manager must show that TSM is good business. Figure 1–6 shows a four-step process that can be used to make this point. The following sections expand on these steps.

Preparation

Preparation involves conducting the research necessary to answer questions such as those shown in Figure 1–7. Notice that only one question in Figure 1–7 (question 5) deals with regulatory compliance, and this question is cast in economic, not compliance, terms. TSM will ensure regulatory compliance, but this is a secondary, not a primary benefit.

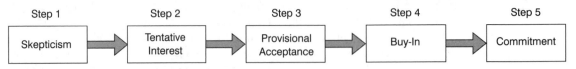

Figure 1–5
Evolution of Commitment

Figure 1–6
Process for Gaining Executive
Commitment

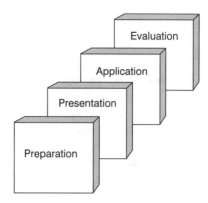

Finding the answers to questions such as those in Figure 1–7 will require the cooperation of the organization's human resources and accounting departments. Consequently, the safety and health manager should begin by securing permission to proceed from the appropriate executive managers. Never begin research without the consent of appropriate executives, including the CEO.

Safety and health managers should make no secret of the fact that they are conducting research. They should make clear that their research is in preparation for a presentation, the goal of which is to gain executive commitment to TSM. Once permission has been received, research can begin. In addition to collecting the desired information for the organization in question, it is a good idea to obtain comparative data from competitors and/or the industry in general.

Presentation

Once the research step has been completed, the next step involves presenting the findings to executive management. The presentation consists of answering the types of questions shown in Figure 1–7. An effective approach is to use visual aids. Figures 1–8 and 1–9 are examples of visual aids that might be used in such a presentation.

The first of these figures shows that the Western Division of the fictitious ABC Company has a safety/health problem. During the past three years the division has lost an average of 74 days of work due to accidents. The company's Eastern Division lost an average of 45 days per year during the same period, and its European Division lost just 25. Figure 1–9 shows that the Western Division also has a productivity problem. It takes this division 13 hours to produce one unit of the company's product. The Eastern Division takes eight hours to produce the same unit, and the European Division takes just six hours. The safety and health manager making this presentation would point out the apparent correlation between accidents and poor productivity.

Figure 1–10 is a checklist of tips that will help enhance the quality and effectiveness of presentations made to executive managers. These are tips that apply to presentations made to small groups—between five and fifteen people—in a conference room or similar setting.

1. Is this company's performance in the marketplace as good as we would like it to be?

2. Does this company have any sustainable competitive advantages over its competitors? If so, what are they?

3. Is this company spending more than it should on workers' compensation costs? Are workers' compensation costs in this company increasing, decreasing, or remaining stable?

4. Are the company's insurance premiums acceptable? Are the premiums increasing, decreasing, or remaining stable?

5. Is product quality as good as we would like?

6. Is product price consistently below that of competitors?

7. Is productivity in this company as high as we would like it to be?

8. Is the number of hours lost due to accidents in this company as low as we would like?

9. Is the absenteeism rate in this company as low as we would like? Is the rate increasing, decreasing, or remaining stable?

10. Is the sick leave utilization rate in this company as low as we would like? Is the rate increasing, decreasing, or remaining stable?

11. How frequently is this company involved in safety/health related litigation? How much is spent annually on this type of litigation? Is the amount spent on safety/health litigation increasing, decreasing, or remaining stable?

12. Is employee morale at this company as high as we would like it to be?

13. Is the employee turnover rate at this company acceptable? Is it increasing, decreasing, or remaining stable?

14. Do employees perceive the work environment in this company as a positive or negative factor in their daily performance?

15. Has this company been required by regulatory agencies to pay safety/health related fines? If so, in what amounts?

Figure 1–7
Questions for Gaining Executive Commitment to TSM

Application and Evaluation

The purpose of the previous step (presentation) was to create enough executive buy-in to allow the TSM implementation to proceed. Full commitment is not likely to occur until executive managers have seen the concept successfully applied. Consequently, it is important at this point to accomplish the following tasks:

- Convince executive management to commit to give TSM their full support long enough to allow for positive results (e.g., six months, one year, 18 months, or what-

Figure 1–8
Lost Days Due to Accidents in
Three Divisions of ABC Company
(Averages of Last Three Years)

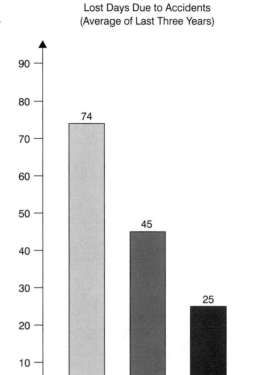

Comparison
Lost Days Due to Accidents
(Average of Last Three Years)

ABC Company

ever makes sense in this situation). Deciding how much time to ask for requires a judgment call on the part of the safety and health manager and will depend on the outcome of the cultural assessment undertaken earlier. TSM might need just six months to catch on in an organization with a favorable culture, but it could take as long as 18 months if the prevailing culture is negative and firmly entrenched. The key is to secure a commitment to give TSM enough time to succeed.

■ Clearly define success and decide how it will be measured. It is important for everyone involved in the implementation of TSM to understand what is expected, what they will be held accountable for, how progress will be assessed, and how performance will be evaluated. The concept of *no surprises* is fundamental to TSM.

With a commitment of sufficient time and a clear charter secured from executive management, application of the TSM implementation model can proceed. As the organi-

Figure 1–9
Productivity Comparison Among
Three Divisions—Hours per Unit
Produced (Averages Last Three
Years)

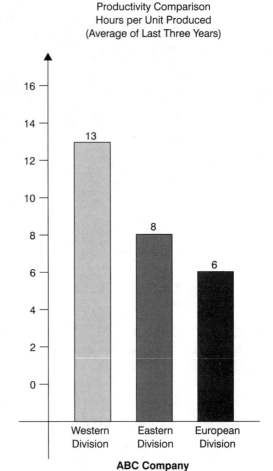

Productivity Comparison
Hours per Unit Produced
(Average of Last Three Years)

zation moves through the various steps in the model, progress will be assessed and performance evaluated. As positive results are achieved, executive commitment will begin to solidify.

MAINTAINING EXECUTIVE COMMITMENT

Initial excitement followed by flagging interest is a common phenomenon when implementing a new concept, particularly when that concept represents major cultural change. Maintaining executive commitment long enough to allow TSM to break the bonds of cultural inertia and become, itself, the cultural norm is a challenge. Figure 1–11 is the Plan-Do-Check-Adjust (PDCA) cycle superimposed on a blanket of continual communication.

- Keep the presentation brief and to the point. Executives are busy people.
- Base the presentation on documented facts. Never make claims you cannot support.
- Use well-designed, attractive visual aids. Keep visual aids simple and make sure all information on them can be easily seen from the back of the room.
- Do at least two complete trial runs (practice sessions) before making the presentation. During the actual presentation is no time to be working out the bugs.
- Arrive early and set up. Test all equipment and have back-up strategies in case something malfunctions during the presentation.
- Make sure the presentation has three distinct components: Introduction, Body, and Summary.
- If you are nervous, concentrate on slowing down your rate of speech. Nervous speakers are prone to rush.
- Make eye contact with all members of the audience, and spread your attention equally. People don't like to be ignored.

Figure 1–10
Presentation Tips for Safety and Health Managers

Every major step in the model for implementing TSM actually consists of a succession of smaller steps that involve applying the PDCA cycle. In each of these steps, the health and safety manager should keep the executive management team informed. What is the plan? Is implementation proceeding on schedule? Are the planned results being produced? What adjustments are being made to counter unforeseen inhibitors?

If the executive management team also serves as the TSM Steering Committee, at each meeting the safety and health manager should give a progress report. If the execu-

---

TSM TIP

Workers' Compensation Categories

Injuries that may be addressed through workers' compensation fall into one of the following categories:

- *Temporary partial disability*
- *Temporary total disability*
- *Permanent partial disability*
- *Permanent total disability*

Figure 1–11
The PDCA Cycle with Continual
Communication

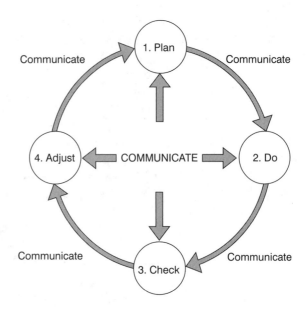

tive management team is not the TSM Steering Committee, it will be necessary to make periodic progress reports to the executive management team during its regularly scheduled meetings. The recommended approach is the oral presentation.

Written reports are easy to file or simply discard without reading. Oral presentations, on the other hand, allow for give-and-take interaction. Periodic face-to-face meetings between the safety and health manager and the executive management team offer several additional benefits for both. These benefits are as follows:

- Opportunities for executive managers to get to know the safety and health manager better and build trust in him or her
- Opportunities for the safety and health manager to build respect and credibility among executive managers
- Opportunities for the safety and health manager to monitor the interest and commitment levels of executive managers
- Opportunities for executive managers to inform the safety and health manager of potential threats and opportunities
- Opportunities for the safety and health manager to clear up the misconceptions and misunderstandings that often accompany cultural change before they undermine executive commitment
- Opportunities for executive managers to suggest strategies the safety and health manager can use to overcome inhibitors
- Opportunities for executive managers to use the meetings as part of the *check* component of their own PDCA cycle

```
┌─────────────────────────────────────────────────────────────────────┐
│                             TSM TIP                                   │
├─────────────────────────────────────────────────────────────────────┤
│                                                                       │
│  Maintaining Executive Commitment                                     │
│                                                                       │
│  One of the most effective ways to maintain commitment to an idea is  │
│  to promise small and produce big. Making big promises and then       │
│  failing to live up to them will cause you to lose both credibility   │
│  and commitment.                                                      │
│                                                                       │
└─────────────────────────────────────────────────────────────────────┘
```

Structure of Periodic Progress Reports

Periodic progress reports given to executive managers should pick up where the presentation made to win their initial commitment left off. The same criteria used to gain executive commitment should be used for maintaining it. If, for example, lost time due to accidents was a criterion in the commitment presentation, it should be a criterion in the progress reports. Has the amount of lost time declined? If so, by how much?

The need for continuity between the criteria used to gain commitment and that used to maintain it underscores the importance of basing the commitment presentation on documented facts. The safety and health manager who overstates during the commitment presentation will pay for it when making periodic progress reports. This can be a sure way to lose commitment.

=========== TSM IN ACTION—STEP 1 ===========

What follows is the first installment of a serialized case study based on the efforts of Mack Parmentier to implement TSM at Meyers Processing Company (MPC). One installment of the case is included with each successive step to illustrate how that step might actually be carried out in a live setting. Although the case is a fictional account provided for the purpose of illustration, it draws from real events that actually occurred.

Mack Parmentier knows he has a tiger by the tail. Meyers Processing Company (MPC) has an abysmal safety and health record. In the previous year, out of 312 work days, only 194 were accident free. Workers' compensation claims were at an all-time high, while productivity and quality were at an all-time low.

MPC is falling behind in the marketplace as its limited resources are increasingly diverted into nonproductive costs associated with accidents and health problems. At the same time the company's best employees are jumping ship, absenteeism is up, productivity is down, quality is unacceptable, and customers are defecting almost daily.

Many of his colleagues had advised Parmentier against accepting the position of Safety and Health Director at MPC. In their words, the job would be a *career killer*. But Parmentier sees the job as an opportunity to try out the TSM philosophy in a situation where it might have an undeniably positive impact. His friends aren't really surprised. For 15 years Mack Parmentier has made a habit of turning lemons into lemonade. As a result, he is well known and highly respected in the safety/health profession. A frequent speaker at professional conferences and a popular consultant in his spare time, Parmentier had been the first among his colleagues to grasp the potential of TSM. Although his colleagues are skeptical about the MPC job, they have confidence in Parmentier. After all, he has helped many of them deal with difficult situations in their own organizations. One of them summed up the feelings of most when he said, "I'm not going to make up my mind too soon. If anybody can turn MPC around, it's Mack Parmentier."

MPC certainly represents a challenge. Parmentier knows that if TSM can be shown to work at MPC, it will more easily gain acceptance in other organizations. Parmentier's assessment of MPC's culture relating to safety and health turned out to be an enlightening experience. His findings speak volumes about why the company's performance is so bad. In fact, in Parmentier's opinion, it's a wonder that things aren't worse. Although he was shocked by some of what the assessment revealed, Parmentier—ever the optimist—sees the situation as an opportunity. He looks forward to helping MPC turn the situation around.

His next step, after completing the cultural assessment, had been to prepare a presentation for MPC's executive management team. The presentation focused on just six key areas: lost time due to accidents, quality, productivity, workers' compensation costs, absenteeism, and employee turnover. The presentation had been well received. In the end, it had accomplished two of Parmentier's highest priority goals. The first goal was to convince TSM's executive managers to give him their support and commitment long enough to let TSM produce results. Agreement among executive managers had been unanimous, although not enthusiastic. Parmentier knew he had been given a clear message: "We're going to give your idea a chance, but you better be right. If TSM doesn't work, you're gone."

The second goal was to convince the executive management team to serve as the TSM Steering Committee. Parmentier had been firm in his insistence on this point. The point he had stressed over and over was that MPC's performance is so bad that turning things around will require the hands-on involvement of the organization's top managers.

After a lengthy and often heated discussion, the executive management team had agreed to serve as the TSM Steering Committee. Parmentier would attend the regularly scheduled meetings of MPC's executive team, leaving meetings once all safety and health issues on the agenda had been dealt with. For convenience, these issues would be placed first on the agenda.

Parmentier is pleased to have gotten MPC off on the right track with its TSM implementation. However, he knows that change is going to come hard, and that every step in the implementation is likely to be an uphill battle. On the other hand, this is the type of battle he knows how to fight. Parmentier looks forward to the next step.

SUMMARY

1. Implementing TSM involves changing an organization's culture. Key elements of an organization's culture are the marketplace, organizational values, organizational role models, and customs, traditions, rites, and rituals.

2. Implementation of TSM requires executive commitment. Nobody in an organization, other than its executive managers, has the authority to build safety and health concerns into the strategic plan, allocate resources for safety and health activities and reward safety and health friendly behavior appropriately.

3. Executive commitment rarely happens all at once. Rather, it evolves over time as the thinking of executive managers goes through the following stages: skepticism, tentative interest, provisional acceptance, buy-in, and commitment.

4. The four-step process for gaining executive commitment is as follows: preparation, presentation, application, and evaluation.

5. Maintaining executive commitment requires establishing realistic expectations, application of the Plan-Do-Check-Adjust model, and constant communication.

KEY TERMS AND CONCEPTS

Application	Organizational culture
Buy-in	Personal involvement
Commitment of resources	Plan-Do-Check-Adjust
Commitment of time	Preparation
Evaluation	Presentation
Executive commitment	Provisional acceptance
Organizational values	Skepticism
Organizational role models	Tentative interest

REVIEW QUESTIONS

1. Explain the following statement: *Implementing TSM will require cultural change in most organizations.*

2. Describe four strategies for promoting cultural change.

3. Why is executive commitment so important in organizations that want to implement TSM?

4. Explain how executive commitment evolves over time.

5. Describe the four-step process for gaining executive commitment.

6. How can executive commitment be maintained once it has been at least tentatively won?

ENDNOTES

1. Peter R. Scholtes, *The Team Handbook* (Madison, Wis.: Joiner Associates, Inc. 1992), pp. 1–16.

2. Terrence E. Deal and Allen A. Kennedy, *Corporate Cultures* (Reading, Mass.: Addison-Wesley, 1982), p. 15.

3. John W. Newstrom and Keith Davis, *Organizational Behavior,* 11th ed. (New York: McGraw-Hill, 1993), p. 58.

4. Deal and Kennedy, pp. 164-169.

Establish the TSM Steering Committee

- Role and Responsibilities of the Steering Committee
- Composition of the Steering Committee
- TSM Facilitator's Role on the Team
- Step 2 in Action

The TSM Steering Committee was introduced in Step 1. Discussion now will explain how to establish the Steering Committee for maximum effectiveness. It also explains the advantages and disadvantages of the options available for establishing the Steering Committee.

ROLE AND RESPONSIBILITIES OF THE STEERING COMMITTEE

The role of the TSM Steering Committee is to ensure that the organization's safety and health-oriented guiding principles are adhered to and that its safety and health objectives are accomplished. The Steering Committee's responsibilities are as follows:

- Making safety and health concerns a part of the organization's strategic plan
- Safety and health policies for the organization
- Oversight of the organization's overall safety and health program
- Approval and disapproval of recommendations from IPTs
- Allocation of the resources needed to support the overall program
- Approval of charters that are drafted by the TSM Facilitator for IPTs
- Making safety and health part of the organization's performance appraisal and reward-recognition system

TSM TIP

TSM Steering Committee Is the Coaching Staff

"The Steering Committee is similar to the coaching staff of a football team. Top executives cannot possibly attend to all the details of a business, make all the decisions, and execute all the functions."[1]

COMPOSITION OF THE STEERING COMMITTEE

The composition of the Steering Committee depends on whether the organization adopts the executive or the delegated approach. With the executive approach, the organization's executive management team serves as the TSM Steering Committee. With the delegated approach, one representative is selected from each department in the organization. In addition, one executive manager serves on the Steering Committee to provide permanent liaison with executive management.

Figure 2–1 is the organizational chart for a fictitious industrial firm. Figure 2–2 is the organizational chart for a fictitious service company. These organizational charts are used in the next two subsections to illustrate the executive and delegated approaches to establishing the TSM Steering Committee.

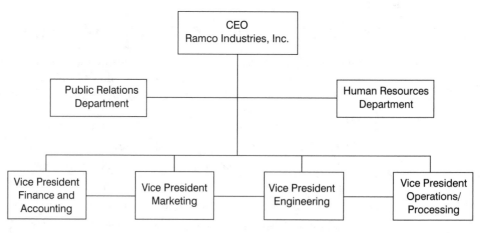

Figure 2–1
Ramco Industries Organizational Chart

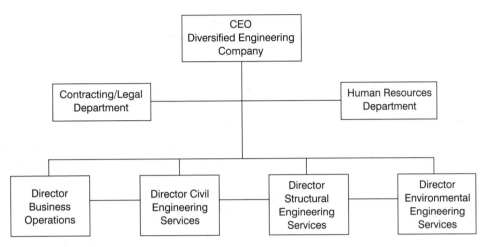

Figure 2–2
Diversified Engineering Company Organizational Chart

════════════ TSM CASE STUDY ════════════

What Would You Do?

Nigel Moore, CEO of Keltron, Inc. intends to move forward with the implementation of TSM. Keltron's profits are being drained at an alarming rate by costs associated with accidents. The company has grown so rapidly that safety and health considerations have been swept aside in the rush to keep up with demand. As a result, Keltron's plant has become a dangerous place.

Moore intends to have his executive council serve as the TSM Steering Committee. He knows that the first question his executives will ask is, "What does this mean to us in terms of time and responsibility?" If you were in Moore's place, how would you answer this question?

Executive Approach

Ramco Industries, Inc. (Figure 2–1) adopted the executive approach in establishing its TSM Steering Committee. The Steering Committee consists of the following members:

■ Chief Executive Officer
■ Vice President for Finance and Accounting

- Vice President for Marketing
- Vice President for Engineering
- Vice President for Operations/Processing
- Director of Human Resources
- Director of Public Relations

The TSM Facilitator is Ramco's Director of Safety and Health who reports to the Vice President for Operations/Processing. At the beginning of the executive management team's weekly staff meeting, the group convenes as the TSM Steering Committee. The TSM Facilitator attends to answer questions, provide updates, and receive assignments. This individual serves as a bridge between executive management and employees on matters of safety and health. Once all safety and health agenda items have been disposed of, the TSM Facilitator leaves and the group resumes its regular role as Ramco's executive management team.

Delegated Approach

Diversified Engineering Company (DEC) (Figure 2–2) adopted the delegated approach in establishing its TSM Steering Committee. The Steering Committee consists of the Director of Environmental Engineering and one representative each from the following departments:

- Contracting/Legal
- Human Resources
- Business Operations
- Civil Engineering Services
- Structural Engineering Services
- Environmental Engineering Services

The TSM Facilitator is a Safety Engineer who works in the Environmental Engineering Services Department. DEC does not maintain a fulltime safety and health staff. Rather, the safety engineer who serves as the TSM Facilitator wears two hats. One hat involves overseeing DEC's internal safety and health program. The other hat involves serving as DEC's resident safety and health expert on environmental-engineering contracts that require this type of expertise.

The TSM Steering Committee meets weekly. It is chaired by the Director of Environmental Engineering. The TSM Facilitator staffs the committee, takes the minutes, monitors assignments made, advises the committee on safety, health, and compliance concerns, and oversees the work of IPTs. The Committee Chair, who also serves on DEC's executive management team, keeps executives apprised of the committee's activities.

TSM FACILITATOR'S ROLE ON THE TEAM

A major factor in the success of the Steering Committee is how well the TSM Facilitator plays his or her role. The role of the facilitator can be difficult. This individual must simultaneously accomplish the following:

- Provide leadership to the Steering Committee without chairing it.
- Guide the Steering Committee in the right direction without pushing it.
- Keep the Steering Committee informed without being perceived as a narrowly focused *know-it-all*.
- Coordinate the Steering Committee's activities without overstepping and assuming the Chair's duties.

Duties of the Facilitator

The TSM Facilitator—whether engineer, technologist, or manager—is the individual within the organization who is responsible to executive management for the organization's overall safety and health program. Regardless of how the TSM Steering Committee is established—executive approach or delegated approach—the Facilitator performs the following duties (Figure 2–3):

- *Staff the Steering Committee.* This duty consists of the following activities: (a) working with the chair to plan the agenda for meetings; (b) taking minutes during meetings; (c) maintaining all applicable records; and (d) undertaking other support duties as assigned by the Steering Committee.
- *Recommend policies.* This duty consists of keeping the Steering Committee informed concerning the need for safety and health related policies. It also involves preparing first drafts of policies for the Steering Committee to consider and adopt.
- *Advise the Steering Committee.* The TSM Facilitator is the Steering Committee's resident expert on all safety and health issues. Consequently, this duty consists of advising the Steering Committee on the technical and compliance aspects of safety and health.

TSM TIP

The TSM Facilitator

"[Such] teams frequently use a designated facilitator—someone who steps aside from the content and concentrates on the process the team is using to do its work."[2]

Figure 2–3
Responsibilities of the TSM
Facilitator

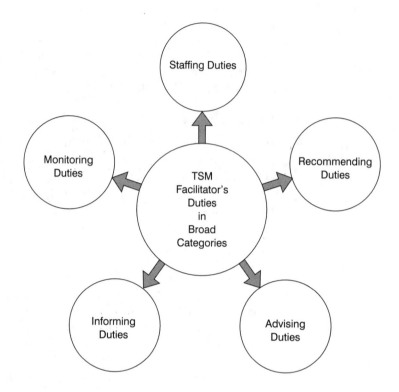

- *Inform the Steering Committee.* This duty consists of keeping the Steering Committee informed concerning the performance of IPTs and the overall performance of the organization in the areas of safety and health.
- *Monitor safety and health concerns.* This duty consists of the following (a) monitoring the performance of IPTs; (b) monitoring the organization's overall performance in the areas of safety and health; and (c) monitoring all applicable compliance requirements. Figure 2–4 is a list of just some of the government organizations that produce safety and health regulations. Although TSM is a performance-oriented concept, it is still important for the TSM Facilitator to keep the organization informed of and in compliance with all applicable regulations.

TSM TIP

Human Factor in Accidents

The human factor theory of accident causation attributes accidents to human error such as overload, inappropriate response, and inappropriate activities.

Figure 2–4
Agencies That Produce Safety
and Health Regulations

> American Public Health Association
>
> Bureau of Labor Statistics
>
> Bureau of National Affairs, Inc.
>
> Commerce Clearing House
>
> Environmental Protectional Agency (EPA)
>
> National Institute for Occupational Safety and Health (NIOSH)
>
> Occupational Safety and Health Administration (OSHA)
>
> Occupational Health and Safety Review Committee (OHSRC)
>
> U.S. Consumer Product Safety Commission

Meeting Facilitation Strategies

An important and challenging aspect of the TSM Facilitator's job is facilitating meetings. Whether the meeting involves the Steering Committee or an IPT, there is a facilitation role to be played, and it is an important role. Hartzler and Henry recommend the following strategies for ensuring effective meeting facilitation:[3]

- *Lead by example.* This is accomplished by adhering to the following rules-of-thumb: (a) participate, but don't dominate; (b) be supportive of diverse views; (c) listen carefully and encourage others to follow suit; and (d) stay positive, don't become defensive.
- *Move the group toward accomplishing the task.* This is facilitated by (a) promoting interaction among group members; (b) keeping the group focused on its task, direction, and agenda; and (c) at key points, summarizing the major points that have been made.
- *Maintain positive team relations.* This is best done in the following ways: (a) identify underlying conflicts and resolve them; (b) make sure every team member's opinion is heard; (c) encourage quiet people to participate, but without embarrassing them; (d) rein in overly talkative members who tend to dominate; (e) promote openness and acceptance; and (f) monitor the reactions of team members.

TSM TIP
Workplace Stress
Workplace stress is caused by an individual's feelings about perceived differences between the demands of the job and his or her ability to cope with the demands.

STEP 2 IN ACTION

Mack Parmentier's first meeting with MPC's TSM Steering Committee had been interesting. The first item on the agenda had been the role and responsibilities of the Steering Committee. Parmentier had taken advantage of the opportunity to explain not just the role of the Steering Committee, but also his role as the TSM Facilitator, and the relationship of the Facilitator and the Steering Committee.

Parmentier had come away from the first Steering Committee meeting with some clearly formed impressions. First, the Steering Committee is far from being a team. Rather, it is a collection of autonomous individuals, each pursuing his own agenda independently of the others. Teambuilding will have to be a high priority. Second, the Steering Committee members are poorly informed concerning safety and health issues. Safety and health awareness training will also have to be a high priority.

However, Parmentier's most pressing challenge will be to develop a positive relationship with MPC's CEO, Jim Farmer. Farmer is not accustomed to working with a facilitator. Consequently, he doesn't yet know how to make the best use of Parmentier in preparing for and conducting meetings. Bringing Farmer along in this regard will require both tact and perseverance. Parmentier will have to lead by suggestion, always a difficult undertaking. Fortunately, Parmentier is a master of patient persistence.

SUMMARY

1. The role of the TSM Steering Committee is to ensure that the organization's safety and health-oriented guiding principles are adhered to and that its safety and health objectives are accomplished.

2. The TSM Steering Committee is composed in one of two ways; the executive approach or the delegated approach.

3. The TSM Facilitator's role is to provide leadership to the Steering Committee without chairing it, guide the Steering Committee without pushing it, keep the Steering Committee informed, and coordinate the Steering Committee's activities.

4. The TSM Facilitator's duties include the following: recommend policies, advise the Steering Committee, inform the Steering Committee, and monitor safety and health concerns.

KEY TERMS AND CONCEPTS

Advising the Steering Committee   Monitoring safety and health concerns

Delegated approach   Recommend policies

Executive approach   Staff the Steering Committee

Informing the Steering Committee   TSM Facilitator

Lead by example   TSM Steering Committee

Maintain positive team relations

REVIEW QUESTIONS

1. Describe the role of the TSM Steering Committee.
2. List four specific responsibilities of the Steering Committee.
3. Explain the difference between the executive and delegated approaches in establishing the TSM Steering Committee.
4. What is the TSM Facilitator's role as it relates to the Steering Committee?
5. What do the following duties of the TSM Facilitator involve?
   - Staffing the Steering Committee
   - Advising the Steering Committee
6. How can the TSM Facilitator maintain positive relations among team members?

ENDNOTES

1. David L. Goetsch and Stanley B. Davis, *Implementing Total Quality* (Upper Saddle River, N.J.: Prentice Hall, 1995), p. 58.
2. Meg Hartzler and Jane E. Henry, *Team Fitness* (Milwaukee, Wis.: ASQC Quality Press, 1994), p. 295.
3. Hartzler and Henry, pp. 235–236.

Mold the Steering Committee into a Team

In traditional organizations committees are often groups of individuals pursuing separate agendas. If TSM is to succeed, the Steering Committee must become a team with all its members working together to move the organization toward a common vision. Step 3 begins the process of turning the Steering Committee into an effective team.

TEAM BUILDING AND THE STEERING COMMITTEE

No matter how the Steering Committee is constituted—whether by executive or delegated approach—its members are likely to come to the table with ingrained departmental loyalties. This is because traditionally their performance has been evaluated based on performance within their subdivision of the overall organization (their division, department unit, team, etc.), and their job security and career advancement have depended on these evaluations. Consequently, members of newly formed teams often begin as individuals who must be turned into team players.

43

TSM TIP
Team Building *"Team building encourages team members to examine how they work together to identify their problems and to develop more effective ways of cooperating. The goal is to make the team more effective. High-performance teams accomplish their tasks, learn how to solve problems, and enjoy satisfying interpersonal relationships with each other."*[1]

An effective approach in undertaking initial team-building activities for the Steering Committee is to hire an outsider who conducts the training off-site. This is especially true when the Steering Committee is the organization's executive management team, but the practice is also recommended in the case of a delegated Steering Committee. In neither case should the TSM Facilitator conduct or participate in the team-building activities. This is because the Facilitator's relationship with the Steering Committee is an arm's-length relationship.

Conducting the initial teamwork training off-site protects against interruptions, thereby allowing members to focus on the issue at hand. Bringing in an outsider helps the team develop faster because it allows individual members to drop their social defenses and open up.

STAGES IN TEAM DEVELOPMENT

People who are accustomed to working as individuals don't become team players all at once. Rather, they typically go through at least four stages. According to Newstrom and Davis, these stages are as follows:[2]

- Forming
- Storming
- Norming
- Performing

Forming

This is the *feeling-each-other-out* stage. As the team is formed, members will be unsure of where they stand and how they fit in. Members will be socially on guard and interaction will be tentative. People tend to respond this way in a new and unfamiliar setting or relationship, and meeting *as a team* will be new and unfamiliar no matter how long the members have worked together as individuals.

Storming

This is the stage in which individual team members will begin to assert themselves, jockey for position, and try to establish themselves within an unstated hierarchy. Debate concerning the direction the team should take may be intense, and tension among members will surface. Executive-level managers who are accustomed to being the "captain" of the team in their respective departments may have trouble becoming just a member of the team on the TSM Steering Committee.

Norming

This is the stage in which group norms begin to supersede individual behavior. Individual members begin to cooperate, competing interests begin to balance out, and the group begins to come together as a team. This doesn't happen all at once, but in tentative starts and stops as trust begins to develop among members.

Performing

This is the stage in which the group finally becomes a team. Individuals work together in a mutually supportive manner, and evaluate their performance based on individual contributions to accomplishing team objectives.

═══════════════ TSM CASE STUDY ═══════════════

What Would You Do?

Karin McCord, who is Johnson, Inc.'s Vice President for Human Resources, is confused. As a member of Johnson's executive management team, she was also a member of the company's new TSM Steering Committee. The Steering Committee's first team-building activity had gone well. Her fellow members had been cordial, if somewhat tentative. But the second session had been a disaster! Cordiality had gone out the window and been replaced with hostility and posturing. McCord had seen a side of her colleagues that she had not seen before. Frankly, their behavior had shocked her. As an advocate of teamwork, McCord had hoped for a different result. If Karen McCord came to you with her concerns, how would you advise her?

CHARACTERISTICS OF EFFECTIVE TEAMS

Effective teams, regardless of their level or makeup, share several characteristics. These common characteristics are as follows (Figure 3–1):

- Supportive environment
- Team player skills

Figure 3–1
Characteristics of Effective Teams

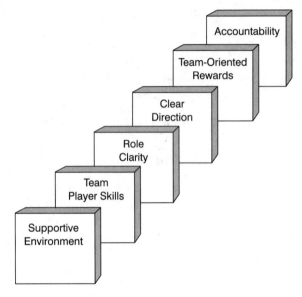

- Role clarity
- Clear direction
- Team-oriented rewards
- Accountability

Supportive Environment

Is the organization structured for teamwork or for individual performance? Does the organization have a team-supportive culture, or does the culture promote individualism? Teams cannot function well if they must function in a nonsupportive environment. A team-supportive environment has the following elements: (a) executive support; (b) cultural expectation of teamwork as the norm; (c) team-based incentives and rewards; and (d) ongoing team-building activities. The characteristics of a team-supportive environment are well known. These characteristics are as follows:

- Open communication
- Constructive, nonhostile interaction
- Mutually supportive approach to work
- Positive, respectful climate

Team Player Skills

Team player skills are personal characteristics of individuals that make them good team players. A study conducted by the Institute for Corporate Competitiveness identified the personal characteristics shown in Figure 3–2 as being traits of good team players. The study concluded that the success or failure of a team will be influenced strongly by the presence or absence of these characteristics in its members.

Role Clarity

On any team, different members play different roles. Consider the example of a football team. When the offensive team is on the field, each of the eleven team members has a specific role to play. The quarterback plays one role, the running backs another, the receivers another, the center another, and the linemen another. Each of these roles is different, but important to the team. When each of these players executes his role effectively, the team performs well.

But what would happen if the center suddenly decided that he wanted to pass the ball? What would happen if one of the linemen suddenly decided that he wanted to run the ball? Of course, chaos would ensue. A team cannot function if team members try to play roles that are assigned to other team members. All members need to understand their respective roles on the team, and play those roles.

Clear Direction

What is the team's purpose? What is the team supposed to do? What are the team's responsibilities? These are the types of questions people have when they are assigned to teams. Such questions should be answered by the team's charter. The various components of a team's charter are as follows:

- *Mission.* The team's mission statement defines its purpose, and how the team fits into the larger organization.

Figure 3–2
Personal Characteristics of
Good Team Players

Institute for Corporate Competitiveness, "Research Study—Competitiveness Factors: Final Report" (1996), pp. 16–21

- Honesty
- Selflessness
- Dependability
- Enthusiasm
- Responsibility
- Cooperativeness
- Initiative
- Patience
- Resourcefulness
- Punctuality
- Tolerance/Sensitivity
- Perseverance

TSM TIP

Team Direction

"Direction is the mechanism that focuses actions purposefully toward team goals. Clear direction helps set priorities that are essential to assigning resources and creates commitment and alignment to the team's purpose."[3]

- *Objectives.* The team's objectives spell out exactly what the team is supposed to accomplish.
- *Accountability measures.* The team's accountability measures spell out how the team's performance will be evaluated.

Figure 3–3 is an example of a team charter for the TSM Steering Committee of a manufacturing company. This charter clearly defines the committee's purpose, where it fits into the overall organization, what it is supposed to accomplish, and how the committee's success will be measured.

Team-Oriented Rewards

One of the most commonly made mistakes in organizations is attempting to establish a teamwork culture while maintaining an individual-based reward system. If teams are to function fully, the organization must adopt team-oriented rewards, incentives, and recognition strategies. For example, teams function best when the financial rewards of its members are tied at least partially to team performance. Performance appraisals that contain criteria in them relating to team performance, in addition to individual performance, promote teamwork. The same concept applies to recognition activities.

Accountability

There is a rule of thumb in management that says, *"If you want to improve performance, measure it."* Accountability is about being held responsible for accomplishing specific objectives or undertaking specific actions. The most effective teams know what they are responsible for and how their success will be measured.

Teams are accountable externally and internally. External accountability concerns the team's responsibility to the overall organization. External accountability measures are spelled out in the team's charter. Internal accountability concerns the responsibilities of individual team members to themselves and to each other (e.g., being honest, dependable, punctual). The most effective teams are held accountable, know what they are accountable for, and know what accountability measures apply.

Figure 3–3
Sample Charter for a TSM Steering Committee

Team Charter
TSM STEERING COMMITTEE
Manufacturing Technologies Corporation

Mission

The purpose of the TSM Steering Committee is to ensure that manufacturing Technologies Corporation has a safe and healthy work environment that is conducive to consistent peak performance. The committee is responsible for the safety and health components of the company's strategic plan.

Objectives

1. To establish and display policies and practices that promote safety and health in the workplace

2. To create and deploy a system through which hazards are identified and eliminated continually.

3. To monitor the effectiveness of the company's hazard-identification-and-elimination system continually.

4. To make the adjustments indicated by the monitoring process.

Accountability Measures

1. Establish benchmarks the company can pursue in the following key areas: lost time due to accidents, workers' compensation costs, absenteeism, productivity, quality, employee perceptions of the work environment, and regulatory fines/costs.

2. Regularly compare actual performance against benchmarks in the following areas: lost time due to accidents, workers' compensation costs, absenteeism, productivity, quality, employee perceptions of the work environment, and regulatory fines/cost.

TSM TIP

Team Accountability

"Accountability is the process of mutually agreeing on what results the team is expected to achieve, specific projects and plans, and how the team will be responsible to the organization and to one another."[4]

POTENTIAL BENEFITS OF TEAMWORK

Teamwork can have both direct and indirect benefits for an organization. Through teamwork, counterproductive internal competition and internal politics are replaced by collaboration. When this happens, the following types of benefits typically accrue:

- Fewer safety and health problems
- Lower turnover
- Less crisis management
- Better quality
- Less conflict among employees
- Fewer interdepartmental barriers
- Better communication organization-wide
- Less rework and waste
- Better morale and employee attitudes
- Better customer satisfaction

The extent to which these benefits are realized will correspond to the effectiveness of an organization's team-building efforts. The better an organization becomes at working in teams, the more of these benefits the organization will enjoy.

TSM TIP

Effective Teams

"Effective teams in action are a pleasure to observe. Members are committed to the firm's success, they share common values regarding safety and customer satisfaction, and they share the responsibility for completing a project."[5]

POTENTIAL PROBLEMS WITH TEAMS

Teamwork can yield important benefits, but as with any concept, there are potential problems. The most pronounced potential problems with teams are as follows;

- It can take a concerted effort over an extended period of time to mold a group into an effective team, but a team can fall apart quickly.
- Personnel changes are common in organizations, but personnel changes can disrupt a team and break down team cohesiveness.
- Participative decision making is inherent in teamwork, but this approach to decision making takes time, and time is often in short supply.
- Poorly motivated and lazy employees can use a team to blend into the crowd, to avoid participation. If one team member sees another slacking, he or she may respond in kind.

These potential problems can be prevented, of course. The first step in doing so is recognizing them. The next step is to ensure that all team members fulfill their responsibilities to the team and to one another.

RESPONSIBILITIES OF TEAM MEMBERS

Internal accountability in teamwork amounts to team members fulfilling their individual responsibilities to the team and to each other. These responsibilities are as follows:[6]

- Interacting with fellow team members in a professional, nonaccusatory, and nondefensive manner
- Being open and honest with fellow team members
- Participating fully in all team activities
- Promoting a sincere desire to work together
- Committing to building the team and fellow team members continually
- Keeping events and conversations that occur within team-building sessions within the group
- Listening to fellow team members
- Promoting understanding among fellow team members

If individual team members fulfill these responsibilities to each other and the team, the potential problems with teams can be overcome, and the benefits of teamwork can be fully realized. It is important for members of the TSM Steering Committee to understand these responsibilities, accept them, and set an example of fulfilling them. If this happens, the benefits to the organization will go well beyond just concerns about safety and health.

ASSESSING NEEDS FOR ADDITIONAL TEAM BUILDING

Team building is an ongoing process that never ends. A periodic needs assessment can help identify additional team-building activities that should be pursued. Conducting periodic needs assessments will help the team focus its ongoing team-building efforts on clearly identified weaknesses. Figure 3–4 is an assessment instrument that can be used by any kind of team, including the TSM Steering Committee and IPTs.

Instructions

To the left of each item is a blank for recording your perception regarding that item. For each item, record your perception of how well it describes your team. Is the statement *Completely True* (CT), *Somewhat True* (ST), *Somewhat False* (SF), or *Completely False* (CF)? Use the following numbers to record your perception.

> CT = 6
> ST = 4
> SF = 2
> CF = 0

Direction and Understanding

_____   1.   The team has a clearly stated mission.

_____   2.   All team members understand the mission.

_____   3.   All team members understand the scope and boundaries of the team's charter.

_____   4.   The team has a set of broad goals that support its mission.

_____   5.   All team members understand the team's goals.

_____   6.   The team has identified specific activities that must be completed in order to accomplish team goals.

_____   7.   All team members understand the specific activities that must be completed in order to accomplish team goals.

_____   8.   All team members understand projected time frames, schedules, and deadlines relating to specific activities.

Characteristics of Team Members

_____   9.   All team members are open and honest with each other all the time.

_____   10.   All team members trust each other.

_____   11.   All team members put the team's mission and goals ahead of their own personal agendas all of the time.

_____   12.   All team members are comfortable that they can depend on each other.

_____   13.   All team members are enthusiastic about accomplishing the team's mission and goals.

_____   14.   All team members are willing to take responsibility for the team's performance.

Figure 3–4
Team-Building Needs Assessment

Using this instrument, team leaders can determine if all members fully understand the team's charter (mission, goals, activities, timetables, etc.). If every member does, in fact, fully understand the team's charter, the team score will be a 6 for each criterion in the Direction and Understanding section of the instrument. To determine the team score for a given criterion, add all of the individual responses from team members and divide by the number of team members. This average score is the team score for the criterion in question. Any team score that is less than 6 indicates a need for team building. For

---

_____ 15. All team members are willing to cooperate in order to get the team's mission accomplished.

_____ 16. All team members will take the initiative in moving the team toward its final destination.

_____ 17. All team members are patient with each other.

_____ 18. All team members are resourceful in finding ways to accomplish the team's mission in spite of difficulties.

_____ 19. All team members are punctual when it comes to team meetings, other team activities, and meeting deadlines.

_____ 20. All team members are tolerant and sensitive to the individual differences of team members.

_____ 21. All team members are willing to persevere when team activities become difficult.

_____ 22. The team has a mutually supportive climate.

_____ 23. All team members are comfortable expressing opinions, pointing out problems, and offering constructive criticism.

_____ 24. All team members support team decisions once they are made.

_____ 25. All team members understand how the team fits into the overall organization/big picture.

Accountability

_____ 26. All team members know how team progress/performance will be measured.

_____ 27. All team members understand how team success is defined.

_____ 28. All team members understand how ineffective team members will be dealt with.

_____ 29. All team members understand how team decisions are made.

_____ 30. All team members know their respective responsibilities.

_____ 31. All team members know the responsibilities of all other team members.

_____ 32. All team members understand their authority within the team and that of all other team members.

_____ 33. All team goals have been prioritized.

_____ 34. All specific activities relating to team goals have been assigned appropriately and given projected completion dates.

_____ 35. All team members know what to do when unforeseen inhibitors impede progress.

example, a team score of 4.2 on item five indicates a need to explain the team's goals more clearly. If the team score for item ten indicates that team members don't completely trust each other, trust-building activities are in order.

Team leaders should never assume that team building is over with and no longer needed. Teams should assess their team-building needs using the instrument in Figure 3–4 or a similar tool at least twice each year.

SUGGESTED TEAM TRAINING PLAN

Figure 3–5 is an outline of a recommended plan for a one-day seminar on team building. This plan assumes that a teamwork trainer—preferably one from outside the organization—will conduct the seminar. The trainer should meet with the TSM Facilitator a day or two before the seminar to rough out drafts of the team's mission, goals, objectives, and accountability measures. The trainer will use these drafts as trial balloons to get conversation started during the applicable parts of the seminar. The drafts are just that, rough versions of these documents. Members of the TSM Steering Committee may accept them, revise them, or even reject them completely.

Even so, it is important to have draft material. Such material can serve two purposes. First, the material will have a catalytic effect in that it will get the ball rolling. Second, draft material gives the TSM Facilitator—a safety and health professional—input into the process.

The recommended approach for molding the TSM Steering Committee into a team is to begin with a seminar such as that suggested in Figure 3–5. The six parts of this seminar are explained as follows:

Rationale for Teamwork Training

The seminar begins with a brief explanation of the rationale for the day's activities. Explaining why the group is going through the training is critical. People have a tendency to think that they already know about teamwork, and that they are already team players. Typically, this is not the case. The point should be made—tactfully, of course—

Figure 3–5
Outline for Initial Team-Building Training

Part #	Training Topic/Issue	Approx. Time
1	Rationale for teamwork training	½ hour
2	Description of the training	½ hour
3	Direction and understanding (Team mission, goals, and objectives)	1 hour
4	Characteristics of team players	1 hour
5	Accountability measures	1 hour
6	Initial team-building action plan	4 hours

that most people have to learn to be team players. Exceptions to the rule who are natural team players can learn to be even better at teamwork.

Description of the Training

This part of the seminar consists of an overview of the day's training. It is important to let participants know what is in store for them during the day. There is an old adage concerning training that says, "Tell them what you are going to tell them. Then tell them. Then tell them what you told them." This part of the seminar satisfies the tell-them-what-you-are-going-to-tell-them component.

Direction and Understanding

This section consists of a brainstorming session during which participants develop the team's mission, goals, and objectives. The trainer serves a facilitating role for this portion of the seminar; his or her job being to draw out all participants and to ensure that individual participants don't dominate.

The starting point for initiating discussion should be the draft material prepared by the TSM Facilitator. This material will break the ice and focus the attention of participants on the task at hand.

Characteristics of Team Players

This is a critical step in the seminar. It is the point at which the topic of personal characteristics is broached. Team players must either have, develop, or adopt such characteristics as the following:

- Cooperativeness
- Selflessness
- Patience
- Punctuality
- Supportiveness
- Honesty
- Enthusiasm
- Perseverance
- Initiative
- Dependability
- Resourcefulness
- Tolerance/sensitivity
- Trustworthiness
- Responsibility

These types of characteristics are not developed in a one-hour component of an eight-hour seminar. They develop over time. However, they can be introduced in a seminar format, discussed, and acknowledged as being important. In addition, the seminar can establish expectations concerning these characteristics.

Accountability Measures

With the team's mission, goals, and objectives in place, participants can decide how they will measure performance and progress. This is the accountability issue, and no aspect of team building is more important. Every member of the team needs to understand how success will be handled. In addition, it is important for team members to understand their own responsibilities and those of their teammates, how team decisions are made, and how unforeseen inhibitors are to be handled.

Initial Team-Building Action Plan

The team's mission, goals, objectives, and accountability measures are pulled together in this component of the seminar. The character traits of team players are examined critically, and a discussion is undertaken that revolves around the following question: What do we need to do to become an effective team?

All team-building needs suggested are recorded on a flip chart by the trainer. Once the list is complete, the items on it are prioritized. An action plan is then developed for satisfying the needs identified. For example, if a need is identified in the area of trust, trust-building activities should be included in the action plan. Figure 3–6 is a team-building action plan developed by a TSM Steering Committee.

Figure 3–6
Sample Action Plan for One
Step of Team Building

**Team-Building Action Plan
TSM Steering Committee
The Berkley Group, Inc.**

- **Trust-Building**
  Schedule an off-site trust-building activity.

- **Accountability**
  Bring in a performance-measurement expert to help improve the team's accountability measures.

- **Facilitator/Team Relationship**
  Schedule a meeting involving the TSM Facilitator to work out the details of the Facilitator's role and relationship with the Steering Committee.

- **Tolerance/Sensitivity**
  Arrange a team-building activity to help team members learn to deal better with diversity issues within the team.

STEP 3 IN ACTION

Mack Parmentier was pleased to learn from the trainer who had conducted the Steering Committee's team-building seminar that all had gone well. The seminar had been conducted in a conference room in a local hotel. Steering committee members had agreed to turn off their pagers and cellular telephones so as to prevent interruptions.

The mission, goals, and objectives Parmentier had drafted for the trainer had been reviewed thoroughly, discussed at length, modified somewhat, and adopted. The accountability measures had given several Steering Committee members a problem. Apparently their problem had more to do with the concept of being held accountable than with the actual accountability measures. One of the participants had asked, "Why should we be accountable for safety and health issues? That's why we have Mack Parmentier and his department."

Jim Farmer—MPC's CEO—had handled this question well. He had made the point that as executive managers, the members of the Steering Committee are responsible for every aspect of MPC's performance, including safety and health. Farmer had summarized his concept of responsibility as follows: "Everyone in this room is a vice president or director. In your departments, you are in charge and everyone in this room would challenge me if I said otherwise. Well folks, you can't be in charge and not be responsible. We are MPC's executive managers, we are in charge, we are responsible, and—therefore—we must be accountable."

SUMMARY

1. Teams develop in stages. These stages are as follows: forming, storming, norming, and performing.

2. Effective teams have the following characteristics: supportive environment, team capabilities, role clarity, clear direction, team-oriented rewards, and accountability.

3. The potential benefits of teamwork are as follows: fewer safety and health problems, lower turnover, less crisis management, better quality, less conflict, fewer interdepartmental barriers, better communication organization-wide, less waste and rework, better morale, better employee attitudes, and better customer satisfaction.

4. Potential problems with teams are as follows: it can take a long time to mold a group into a team; personnel changes can disrupt a team; team-based decision making takes longer; and lazy employees can use the team to blend in to escape involvement.

5. Responsibilities of team members are as follows: interacting professionally, being open and honest, participating fully, promoting a desire to work together, committing to building the team continually, maintaining team confidentiality, listening to fellow team members, and promoting understanding.

6. The following aspects of team building should be assessed: direction and understanding, characteristics of team players, and accountability.

7. The suggested team training plan for the TSM Steering Committee contains the following components: rationale for teamwork training, description of the training, direction and understanding, characteristics of team players, accountability, and initial team-building action plans.

KEY TERMS AND CONCEPTS

Accountability

Characteristics of team players

Clear direction

Direction and understanding

Forming

Norming

Performing

Role clarity

Storming

Supportive environment

Team capabilities

Team-oriented rewards

REVIEW QUESTIONS

1. Why is it so important to mold the TSM Steering Committee into a team?
2. Explain the four stages of team building.
3. List the characteristics of effective teams.
4. Explain the potential problems with teams.
5. What are the responsibilities of team members?
6. Describe how to assess needs for team-building activities.
7. Outline an initial teamwork training plan for a hypothetical steering committee.

ENDNOTES

1. John W. Newstrom and Keith Davis, *Organizational Behavior: Human Behavior at Work*, 9th ed. (New York: McGraw-Hill, 1993), p. 307.
2. Newstrom and Davis, pp. 425–426.
3. Meg Hartzler and Jane E. Henry, *Team Fitness* (Milwaukee, Wis.: ASQC Quality Press, 1994), p. 5.
4. Hartzler and Henry, p. 7.
5. Newstrom and Davis, p. 428.
6. David L. Goetsch and Stanley B. Davis, *Implementing Total Quality* (Upper Saddle River, N.J.: Prentice Hall, 1995), p. 70.

Give the TSM Steering Committee Safety and Health Awareness Training

Once the Steering Committee has completed its teamwork training and initial team-building activities, it is ready for Step 4. This step is intended to make members sufficiently knowledgeable of safety and health concerns to effectively fulfill their roles on the TSM Steering Committee.

RATIONALE FOR TRAINING THE STEERING COMMITTEE

Regardless of whether the organization adopts the executive or the delegated approach in establishing the TSM Steering Committee, the committee members will not be safety and health experts, nor do they need to be. But they do need to be aware of how safety and health affect organizational performance. Figure 4–1 shows the four most important reasons for giving the TSM Steering Committee safety and health awareness training.

Figure 4–1
Why Train the Steering
Committee?

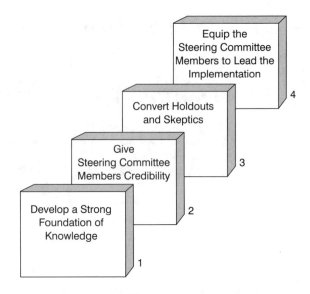

Develop a Foundation of Awareness

Since the Steering Committee will guide the organization in the implementation of TSM, it is important for members to understand how safety and health affect performance. In addition, committee members need to be aware of the most important safety and health issues affecting the organization.

Give Steering Committee Members Credibility

If TSM Steering Committee members are going to lead the implementation, they must have credibility. Speaking the language of safety and health will build credibility. Consequently, the awareness training is designed to help committee members begin to speak the language of TSM, and to have a better understanding of safety and health issues.

Convert Skeptics

TSM will not be welcomed by all executive managers in an organization, at least not at first. Some managers will have questions, and others will have misgivings. The awareness training will give skeptics an opportunity to ask questions and state their misgivings. Often when people have an opportunity to express their misgivings in an open, nonjudgmental environment, their skepticism can be overcome.

Equip Steering Committee Members to Lead

Awareness training should equip members of the Steering Committee to lead the organization through the implementation of TSM. During the training, committee members

---

TSM TIP

Training the Steering Committee

"Training must take place at all levels within the organization. However, it must start at the top. Management must be the driving force behind the transformation process. Everything can improve—even management."[1]

---

should learn enough about safety, health, and TSM to be able to speak the language, understand the problems, and overcome the barriers.

WHO CONDUCTS THE TRAINING?

Awareness training should be conducted by a safety and health professional. Organizations have two options when deciding who should conduct the training. The first option is to use the TSM Facilitator. The second option is to bring in an outside consultant.

The preferred approach is to use the TSM Facilitator. This individual is a safety and health professional. Consequently, he or she has the necessary technical expertise. Using the TSM Facilitator as the trainer gives this individual an opportunity to build relationships with the Steering Committee members, and it lets the Facilitator build credibility in the eyes of committee members.

The only reason for bringing in an outside consultant would be that members of the Steering Committee might not be comfortable with the idea of being trained by a subordinate. If this option is exercised, the TSM Facilitator should meet with the consultant to plan the training.

SAFETY AND HEALTH AWARENESS CURRICULUM

No attempt is made to convert Steering Committee members into safety and health experts. It isn't necessary that they be experts. Rather, the members need to be aware of safety and health concerns, and know enough to be credible participants in helping create and maintain a work environment that is conducive to consistent peak performance. Figure 4–2 is an outline for a safety and health awareness seminar designed especially for TSM Steering Committee members. The seminar consists of a minimum of four hours of training spread comfortably over one day. The recommended schedule has an hour for lunch, three 15-minute breaks, and a 30-minute wrap-up session. This approach gives the trainer room to accommodate going over time without encroaching on each successive session. The various seminar topics are explained in the following sections of this discussion.

Figure 4–2

Sample Coverage for Safety and
Health Awareness Seminar

Seminar Outline Safety and Health Awareness	
Topic	**Time**
1. Accidents and their effects	45 minutes
2. Safety, health, and competitiveness	1 hour
3. Safety analysis/hazard prevention	45 minutes
4. Promoting safety and health	45 minutes
5. Ergonomics and safety	45 minutes
Total	**4 hours**
Recommended Schedule	
9:00 a.m.	**Topic 1**
9:45 a.m.–10:00 a.m.	Break
10:00 a.m.–11:00 a.m.	**Topic 2**
11:00 a.m.–12 Noon	Lunch
12:15 p.m.–1:00 p.m.	**Topic 3**
1:00 p.m.–1:15 p.m.	Break
1:15 p.m.–2:00 p.m.	**Topic 4**
2:00 p.m.–2:15 p.m.	Break
2:15 p.m.–3:00 p.m.	**Topic 5**
3:00 p.m.–3:30 p.m.	Wrap-Up

Accidents and Their Effects

This portion of the seminar makes participants aware of the various negative effects of accidents in the workplace. It is important that members of the TSM Steering Committee understand how an unsafe work environment can affect the organization's ability to compete. Some of the information presented during this session is based on national figures. These figures will help participants appreciate the big picture. Correspondingly, some of the information is specific to the organization in question. These figures will localize the issue and give it more gravity.

========== TSM CASE STUDY ==========

What Would You Do?

Mark Dunn is the TSM Facilitator and Safety Manager for MicroEngineering Corporation. The next step in the company's TSM implementation plan involves safety and health awareness training for the TSM Steering Committee. Everything is in order, but there is a glitch. MicroEngineering's TSM Steering Committee members are not convinced that they need the training. Dunn has been asked to meet with the Steering Committee to explain why awareness training is necessary. If you were Mark Dunn, how would you explain the need to the Steering Committee?

This session covers the following topics:

- Cost of accidents (national and local)
- Lost time due to injuries (national and local)

We will now look at them in more detail.

Cost of Accidents

When presenting this topic the trainer should begin with national figures such as those shown in Figure 4–3. The national figures should be followed by figures in the same or similar categories for the organization, as illustrated by Figure 4–4. Providing figures specific to the organization will require some in-house research. This is one of the reasons the TSM Facilitator is recommended as the trainer. He or she has or can gain access to the records needed to identify and summarize local costs. An outside consultant will not have the necessary access. An outside consultant, if used, will have to work closely with the TSM Facilitator in preparing seminar materials.

Figure 4–3
Accident Costs by Categories
(Typical Annual Figures)

Lost wages	$38 billion
Medical costs	$24 billion
Insurance administration	$28 billion
Property damage	$27 billion
Fire Losses	$9 billion
Indirect costs	$23 billion

Figure 4–4
EMCO PRODUCTS, INC.
Accident Costs by Categories
(Previous Year)

Lost wages	$474,365
Medical costs	$379,561
Insurance administration	$272,111
Property damage	$294,876
Fire Losses	$37,693
Indirect costs	$194,322

TSM TIP

Cost of Accidents

"Clearly accidents on and off the job cost U.S. industry dearly. Every dollar that is spent responding to accidents is a dollar that could have been reinvested in modernization, research and development, facility upgrades, and other competitiveness enhancing activities."[2]

Lost Time Due to Injuries

Businesses in the United States lose more than 35 million hours per year as a result of accidents. This figure is direct time lost because of disabling injuries. It does not include additional time lost to periodic medical examinations that may be necessary after the employee returns to work.

Once participants are familiar with the lost-time issue at the national level they should be shown the figures for their organization. Figure 4–5 is an example of a visual aid that might be used for conveying the lost-time costs of a specific company.

EMCO Products, Inc. maintains a total workforce of 156 employees. Without overtime, these employees work a minimum of 2,000 hours per year each, or 312,000 total. EMCO loses 12 percent of these hours, or 37,440 hours per year due to injuries. When multiplied times EMCO's average loaded labor rate ($12.67 per hour), the company is shown to lose almost a half million dollars per year in lost time due to injuries. This is the type of information that will drive home the importance of a safe and healthy workplace to Steering Committee members.

Safety, Health, and Competitiveness

It is important for all members of the TSM Steering Committee to understand that safety and health can improve their organization's competitiveness. In order to survive and succeed in today's global marketplace, organizations must be competitive. Companies that used to compete only on a regional basis now find themselves competing against companies from the Pacific Rim, Europe, South America, and Canada.

In my book *Occupational Safety and Health* (2nd ed., Prentice Hall, 1996), I describe the effects competition can have on safety and health in the following terms:

Figure 4–5
Cost Summary
Lost Time Due to Injuries at
EMCO Products, Inc.

Employee Hours Lost	×	Average Loaded Labor Rate	=	Approximate Cost
37,440	×	12.67	=	$474,364.80

The need to achieve peak performance levels day after day puts intense pressure on companies, and pressure runs downhill. This means it is felt by all employees from executive-level managers to workers on the shop floor. It is not uncommon for this pressure to create a harried atmosphere that can increase the likelihood of accidents. It can also lead to shortcuts that increase the potential for health hazards (e.g., improper storage, handling, and use of hazardous materials). This is unfortunate because the most competitive companies typically are also the safest and healthiest.[3]

Competitiveness, as it relates to an organization, is the ability to consistently succeed in the marketplace. Figure 4–6 shows the principle factors that, taken together, make an organization competitive. This portion of the safety awareness seminar should be an open, frank discussion of how each of these factors can be affected by the quality of the work environment. Key discussion points are covered next.

Quality

In order to consistently produce a quality product or operate a quality process, employees must be able to move about the workplace freely and without fear of accidents and incidents. For example, employees who must constantly step over or around scattered obstacles, or who must tread cautiously on a slippery floor cannot move freely, nor can they focus 100 percent on maintaining and improving quality.

Figure 4–6
Factors Determining
Competitiveness

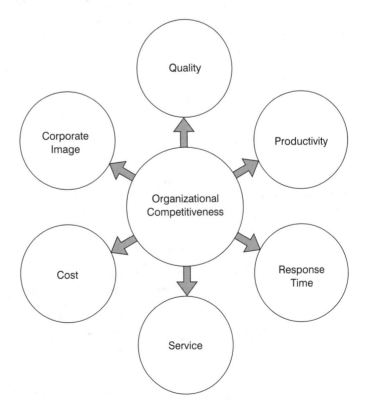

Productivity

Productivity is a function of output versus input. The key is to produce the most output with the least input.

$$\frac{\text{Output}}{\text{Input}} = \text{Productivity}$$

The input element of the formula in a work setting can be thought of as the *Four M's*:

Men (Employees)

Material

Money

Motivation

All four of these standard input factors are affected by the work environment. The safer the work environment, the less likely it is that an employee with valuable expertise will be lost even temporarily because of an accident. When experienced employees who are able to perform their jobs efficiently and effectively must—because of accidents—be replaced by less experienced employees, productivity suffers. Less skilled, less experienced replacement employees typically waste more time and materials than experienced, skilled employees. This waste, in turn, translates into money (increased costs). This increases the amount of input necessary to achieve the same level of output, and as a consequence, productivity suffers. Motivated employees are typically more productive than unmotivated employees. One of the key motivations for employees is the quality of their work environment. This was one of the principle findings of a study conducted by the Institute for Corporate Competitiveness.[4] In this study, 92 percent of respondents listed *quality of the work environment* as being critical in terms of motivation and performance.

Response Time

Response time is the amount of time that elapses between when a customer requests a product or service and when the request is fulfilled. It is a function of people, technology, and management strategies. One of the key determinants of response time is the ability of employees to do their jobs both efficiently and effectively. The faster the response time, the more competitive the organization.

TSM TIP

Reducing Workplace Stress

Managers can reduce the levels of stress experienced by employees on the job by increasing opportunities for feedback, reducing role ambiguity, and increasing job autonomy.

Modern society demands *instant results*. What is wanted, is wanted now. The fast-food industry is built on the concept of response time. The Federal Express success story can be attributed to two things: response time and dependability. A competitive response time is achieved by keeping the best employees working with the best and safest technologies under conditions that encourage both speed and accuracy. An essential management strategy for creating and maintaining these conditions is to provide a safe and healthy work environment.

Service

Service is a critical element in the formula for competitiveness. Consider this factor on a personal level. Is service important to you when eating in a restaurant, having your car repaired, or shopping for clothing? Of course it is. Just as you do, organizations in the marketplace want and expect good service. Organizations that hope to prosper in a competitive environment must provide good service.

Cost

Cost is a critical element of the formula for competitiveness in the global marketplace. One of the most commonly used organizational strategies is seeking to be the low-cost provider of a given product or service in a given industry. According to Thompson and Strickland, "The competitive power of low-cost leadership is greatest when rival products are essentially identical, price competition dominates, most buyers use the product similarly and want similar features, buyer switching costs are low, and large customers shop aggressively for the best price."[5]

Being the low-cost provider in a given industry is a matter of being able to control the costs of all activities in the various components in the *activity/cost chain*, Figure 4–7. The activity/cost chain for a given enterprise consists of all the activities necessary to produce a product and get it to the designated market, or to provide a service. Figure 4–7 is an example of a cost/activity chain for a manufacturing company. Unsafe working

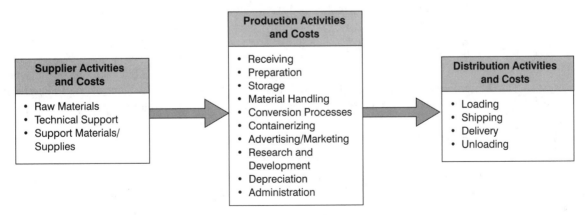

Figure 4–7
Activity/Cost Chain

conditions in an organization can lead to absenteeism, tardiness, and poor productivity. Lost time due to accidents including wages lost, medical expenses, insurance administration, and indirect costs add more than $112 billion to the cost of products produced in the United States every year. Clearly, unsafe working conditions increase the costs associated with production activities which, in turn, increases the cost of an organization's products/services.

Corporate Image

Corporate image is critical in a competitive global marketplace. An image as a safe and healthy employer helps an organization attract and retain the best employees. An image as a producer of safe products helps attract and retain customers. It also makes an organization a less inviting target for enterprising product-liability lawyers. An image as a good corporate citizen concerned about the environment makes an organization a welcome neighbor in a community. The tragedy that occurred in Bhopal, India in 1984 provides a stark illustration of how safety and health issues can affect an organization's corporate image. Poisonous gas leaking from a Union Carbide pesticide plant killed more than 2,000 people and injured approximately 20,000 others. The Indian Supreme Court eventually ordered Union Carbide to pay $470 million dollars to the victims. Although Union Carbide claimed that its plant had been sabotaged by disgruntled employees, negative publicity was carried in the mass media for more than five years following the accident. For Union Carbide, the Bhopal disaster became a corporate-image nightmare that wouldn't go away.

Overview of Safety Analysis and Hazard Prevention

Safety analysis and hazard prevention can be undertaken on a preliminary or a detailed basis. Formal, detailed hazard analysis is the job of safety and health professionals who have the specialized training needed to use such tools as Failure Mode and Effects Analysis (FMEA), Hazard and Operability Review (HZOP), Technic of Operations Review (TOR), Human Error Analysis (HEA), Fault Tree Analysis (FTA), and Risk Analysis. There should be no attempt made to teach members of the TSM Steering Committee how to use these tools. Rather, this component of the awareness training is to help Steering Committee members understand what the results produced by these tools mean.

In addition to interpreting the results of safety analysis tools, Steering Committee members should learn how to systematically control hazards. Each of the following strategies should be presented during this component of the training:[6]

- Eliminate the source of the hazard
- Substitute a less hazardous equivalent
- Reduce the hazard at the source
- Remove the employee from the hazard (e.g., substitute robot/automated equipment)
- Isolate the hazard by enclosing it in a barrier or by some other method
- Dilute the hazard (e.g., ventilate the hazardous substance)

■ Apply appropriate management strategies (e.g., limiting access to the hazard)
■ Use appropriate personal protective equipment
■ Provide employee training
■ Practice good housekeeping

Steering Committee members should also learn the steps for implementing the hazard control measures shown in Figure 4–8. Using the hazard control process shown in this figure, once a hazard has been identified one of the standard strategies listed above is selected and implemented. Safety and health personnel and members of the IPT involved monitor the situation and assess the effectiveness of the chosen strategy. The findings are shared with the Steering Committee along with recommendations for adjustments when adjustments are necessary. It is critical that the Steering Committee be involved in this way because strategy implementation and adjustments to the strategy may cost money or necessitate policy changes, both of which will require executive-level approval.

Promoting Safety and Health

It is important for members of the TSM Steering Committee to understand the critical role they play in promoting safety and health in a TSM setting. That role involves estab-

Figure 4–8
Implementing Hazard Controls

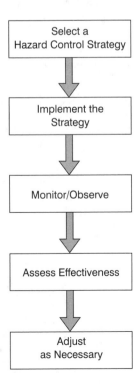

lishing organizational policies for safety and health, and approving the corresponding regulations that translate these policies into everyday practice.

The first lesson Steering Committee members need to learn concerning the promotion of safety and health is that the organization should have a written, published safety policy (see Figure 4–9 for an example). The policy should convey the following distinct messages as a minimum:

- The organization in question is committed to safety and health.
- Employees are expected to perform their duties in a safe and healthy manner.
- The organization's commitment extends beyond the walls of its facilities.

Once an appropriate policy has been drafted, rules and regulations that translate the policy into everyday behavior are needed. Regulations are drafted by safety and health professionals with assistance from employees. However, all regulations drafted should be approved by the Steering Committee.

Ergonomics and Safety

This is a critical component of the Steering Committee's awareness training. Members will probably understand mechanical, electrical, heat, pressure, impact, radiation, fire, and noise-related hazards, but they may not understand that some of the most dangerous hazards are in the category of ergonomics. Ergonomics is the science of fitting the workplace and all of its various elements to the worker. The history of business and industry has been one of expecting the worker to fit into the workplace as best he or she can, often with debilitating results.

Employees whose back injuries develop over time from working hour after hour, day after day in a poorly designed work station are just as injured as their counterparts who hurt their backs suddenly in an accident. Whereas injuries from accidents are sudden, injuries associated with ergonomics tend to develop over time from repeated motion or repeated exposure. Such injuries are known as *cumulative trauma disorders* (CTDs). The Steering Committee's awareness training should cover two broad areas as a minimum: (a) ergonomic risk factors, and (b) ergonomic hazard prevention.

Figure 4–9
Sample Organizational Safety Policy

Monterey Processing Company
Corporate Safety Policy
It is the policy of Monterey Processing Company to provide a safe and healthy work environment for employees. Monterey Processing Company is committed to safety on and off the job. Employees at all levels are expected to conduct themselves with this commitment in mind.

TSM TIP

Ergonomics and Safety

Ergonomics is the science of conforming the workplace to the employee. Common indicators of ergonomic problems include the following: cumulative trauma disorders, absenteeism, high turnover rates, and employee complaints.

Ergonomic Risk Factors

Ergonomic risk factors are conditions that, if left uncorrected, can lead to the development of cumulative trauma disorders. The most common ergonomically hazardous situations are as follows:

- Seated repetitive work (keyboarding, using scissors, etc.)
- Standing work of any kind
- Work involving heavy lifting or repetitive material handling
- Work with hands above the chest
- Work with hand tools

Ergonomic Hazard Prevention

The Occupational Safety and Health Act (OSHAct) contains guidelines for the prevention of ergonomic hazards. In order of priority, ergonomic hazard prevention strategies are as follows:

- Engineering controls
- Work practice controls
- Personal protective equipment
- Administrative controls

Engineering controls involve the proper design or redesign of workstations and tools. Redesigning a work station so that it fits a specific worker instead of the average worker is an example of an engineering control. Providing tools for both left- and right-handed employees is another example of an engineering control. Work practice controls involve the proper design or redesign of the methods used to accomplish work tasks. Changing work methods from a production-line orientation to a work-cell orientation to eliminate constant repetitive motion is an example of a work practice control. Requiring personal protective equipment is another form of control. Administrative controls involve developing work rules and regulations that limit ergonomic hazards. For example, computer operators might be required to take a ten-minute break away from their display terminals every hour. A required break is an administrative control.

STEP 4 IN ACTION

Looking back, Mack Parmentier was pleased with the TSM Steering Committee's safety awareness training. Things had gotten off to a rocky start when several of the steering committee members questioned the necessity of undergoing the training. One vice president had expressed his doubts about the necessity of the training as follows: "It takes all I can do just to keep up in my own field. Now you're asking me to become a safety expert. If we're all going to be safety experts, why do we need Parmentier and his department?" Parmentier had been ready for just such a question, but before he could deliver his well-rehearsed response another vice-president did an excellent job of distinguishing between *awareness* and *expertise*.

This explanation got things rolling on the right track, and Parmentier had been able to move right into a discussion of accidents and their effects. This, and all of the various topics covered during the day-long seminar, had elicited a strong response from participants. At first some of the discussion leaned toward the negative, but after just an hour, participants warmed to the topic and the discussion began to take on a more positive tone. By the end of the day, Parmentier could see that several participants had actually become safety and health advocates.

The portion of the seminar that had really gotten the attention of participants was the *competitiveness* component. Parmentier had made a mental note that "these executives can probably deal with any issue if it is cast in terms of its impact on the company's competitiveness." He had vowed at the time to remember this and use it as a rule of thumb as the TSM implementation proceeded.

SUMMARY

1. The rationale for providing awareness training for the Steering Committee is to: (a) develop a foundation of awareness; (b) give Steering Committee members credibility; (c) convert skeptics; and (d) equip Steering Committee members to lead the implementation of TSM.

2. Awareness training for the Steering Committee may be provided by either the TSM Facilitator or an outside consultant. The preferred approach is to use the TSM Facilitator.

3. The curriculum for the Steering Committee's awareness training should have the following components: accidents and their effects on organizations; health, safety, and competitiveness; safety analysis and hazard prevention; promoting safety and health; and ergonomics and safety.

KEY TERMS AND CONCEPTS

Accidents and their effects

Corporate image

Cost

Ergonomics and safety

Productivity

Promoting safety and health

Quality   Safety analysis and hazard prevention
Response time   Service
Safety, health, and competitiveness

REVIEW QUESTIONS

1. Explain the rationale for providing awareness training for the Steering Committee.

2. Why is it advantageous to have the TSM Facilitator conduct the awareness training for the Steering Committee?

3. What should be covered in the *accidents and their effects* portion of the awareness training?

4. Explain the effect of the work environment on each of the following competitiveness factors as though you were the TSM Facilitator presenting awareness training for the Steering Committee.

   - Quality
   - Productivity
   - Response time
   - Service
   - Cost
   - Corporate image

5. Explain the steps for implementing hazard control measures.

6. What are the most common ergonomically hazardous situations that an organization might confront?

ENDNOTES

1. Stephen Uselac, *Zen Leadership: The Human Side of Total Quality Team Management* (Loudonville, Oh.: Mohican Publishing Company, 1993) p. 101.

2. David L. Goetsch, *Occupational Safety and Health in the Age of High Technology*, 2nd ed. (Upper Saddle River, N.J.: Prentice Hall, 1996), p. 21.

3. Goetsch, p. 341.

4. Institute for Corporate Competitiveness, *Employee Perceptions: Impact of Work Factors on Job Performance*, Report 95-6, August 1995.

5. Arthur A.Thompson, Jr. and A. J. Strickland, III, *Strategic Management: Concepts and Cases*, 7th ed., (Boston: Irwin, 1993), p. 106.

6. Society of Manufacturing Engineers, *Tool and Manufacturing Engineers Handbook* (1994), p. 156.

Develop the Organization's Safety and Health Vision and Guiding Principles

Well-run organizations develop a comprehensive strategic plan and then strive to execute it. In the best organizations the strategic plan is a living, working document that guides all activities at all levels at all times. An organization's strategic plan should contain either a broad objective or a guiding principle, or both, relating to safety and health. This broad objective or guiding principle then becomes the basis for developing the organization's plan for safety and health. The safety and health plan is a subcomponent of the organization's strategic plan that deals specifically with safety and health issues. The first two components of the plan are the organization's safety and health vision and its guiding principles.

SAFETY AND HEALTH VISION DEFINED

An organization's vision describes the dream of what it would like to be from a long-range perspective. It represents an ideal or a beacon in the distance toward which the organization is constantly moving. All activities, decisions, solutions, and expenditures are judged on the basis of how well they advance the organization toward its vision.

The safety and health vision serves the same purpose for the organization except that its scope is more specific. It should be a safety and health beacon in the distance toward which the organization is constantly moving. What follows is an example of a safety and health vision for a modern organization:

The Douglas Corporation will provide an accident-free work environment that is conducive to productivity, quality, and competitiveness; hazard-free products; and environmentally friendly processes.

PURPOSE OF THE SAFETY AND HEALTH VISION

The safety and health vision serves as the guiding force in an organization for ensuring a safe and healthy work environment. It shows not just that management is committed, but what specifically management is committed to. In the example of the Douglas Corporation's vision, the organization showed all stakeholders that it is committed to safety and health by writing down and publishing its safety and health vision. By reading this vision, stakeholders can see that the Douglas Corporation is committed to safety and health in the workplace, to hazard-free products for its customers, and to environmentally friendly processes for its neighbors.

These specific elements of the safety and health vision say much about the Douglas Corporation. To stockholders the vision says, "We have a responsible management team in place that is committed to competing globally and is not likely to surprise us with profit-draining safety and health-related lawsuits." To employees the vision says, "We are concerned about your safety and we understand that you are more likely to perform at peak levels in a safe and healthy environment." To customers the vision says, "We are concerned about your safety and will endeavor to provide you with a product that is hazard-free." Finally, to the community the vision says, "We want to be a good corporate neighbor that is attentive to environmental issues."

This brief portion of the Declaration of Independence paints a word picture of the dream the founding fathers had for the United States. It represents a philosophical ideal

---

TSM TIP

A Powerful Vision

Perhaps the best way to understand what a vision does for an organization is to consider the fact that the Declaration of Independence is the vision for the United States. Consider these stirring words: "We hold these Truths to be self-evident, that all Men are created equal, that they are endowed by their Creator with certain unalienable Rights, that among these are Life, Liberty, and the Pursuit of Happiness. That to secure these Rights, Governments are instituted among Men, deriving their just Powers from the Consent of the Governed. . . . "

that is a beacon by which the United States is guided. At any point in its history that the United States has not lived up to this vision, its citizens have been able to point to the Declaration and ask, "Why?" Without this vision, our country might still be mired in the moral quagmire of slavery, women might still be disenfranchised, and the door to opportunity might still be closed for whole segments of the population. But such practices are difficult to rationalize and perpetuate in the face of a clearly stated vision to the contrary. An organization's safety and health vision serves the same purpose. When an organization with a clearly stated safety and health vision allows unsafe practices to creep in, stakeholders can point to the vision and ask "Why?"

CHARACTERISTICS OF THE SAFETY AND HEALTH VISION

Figure 5–1 shows the characteristics of a well-written safety and health vision. They are as follows:

■ Easily understood by all stakeholders
■ Briefly stated, yet clear and comprehensive in meaning

Figure 5–1
Characteristics of a Well-Written
Safety and Health Vision

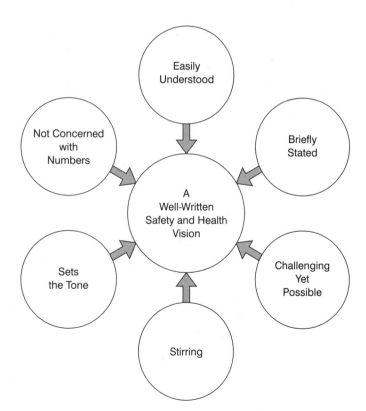

- Challenging, yet possible to accomplish
- Stirring enough to move employees toward unity of purpose
- Sets the tone for how the organization approaches safety and health
- Not concerned with numbers

Easily Understood

The safety and health vision is for all people who have a stake in the performance of the organization. These *stakeholders* include owners (stockholders), all employees at all levels and their families, customers, suppliers, and the community at large. All of these stakeholders benefit either directly or indirectly when the organization in question lives up to its safety and health vision.

Because the vision speaks to a broad and diverse audience, it is important to ensure that it can be easily understood. This means it should be free of *tech-speak* and other forms of organization-specific language. Any person with an interest in the organization should be able to read the safety and health vision and understand what it means.

Briefly Stated

Even the largest corporation should be able to state its vision in no more than a paragraph. Visions that are too long and too wordy are difficult to understand. A good rule of thumb for organizations to follow in developing their safety and health visions is as follows: *Keep the vision short enough that every employee can easily memorize it.* Every employee in an organization should be able to recite the safety and health vision from memory.

The challenge in keeping the vision short is that it must also be comprehensive. This means it must contain enough information to clearly convey the organization's dream with regard to safety and health. The following vision statement was used earlier as an example:

> *The Douglas Corporation will provide an accident-free work environment that is conducive to productivity, quality, and competitiveness; hazard-free products; and environmentally friendly processes.*

TSM TIP

The Vision Statement Must Be Crafted Carefully

The vision statement must be crafted in such a way that all employees can relate to it, and, in so doing, execute their work in a manner that is consistent with its meaning and objectives.[1]

This statement is brief but comprehensive. The entire safety and health vision is stated in just one sentence, yet from this sentence it is clear that The Douglas Corporation is committed to safety and health for its employees, customers, and neighbors. This is an example of what is meant by brief but comprehensive.

Challenging Yet Possible

A good safety and health vision will challenge the organization and its employees. It should cause the organization to stretch, but not so much that the vision is seen as being unrealistic. For example a vision that says, "ABC Corporation will lose no more than 10 hours per day to accidents," would not be challenging. An organization could most likely live up to this vision without even trying. Such a vision would convey the message that ABC Corporation is not really committed to safety and health.

On the other hand, a vision that says, "There will never again be an accident at XYZ Corporation," probably goes too far in the other direction. Employees are likely to see such a vision as being impossible and just tune it out. Only by achieving the proper balance in developing the vision can an organization properly challenge its employees.

Stirring

Organizations need all stakeholders—but especially employees—to unite around the vision, and make it their own. Consequently, a vision must appeal to employees on an emotional level. It must represent something that is lofty and good; something that is important enough to justify their interest and efforts; something they are proud to strive for. The Douglas Corporation's vision satisfies this criteria. "An accident-free-work environment that is conducive to productivity, quality, and competitiveness" is certainly an ideal around which employees at all levels can unite. Making a commitment to produce "hazard-free products" is certainly something that is lofty and good. Finally, being a good neighbor in the community is something of which all employees can be proud.

Sets the Tone

A well-written vision sets the tone for how the organization will conduct itself with regard to the work environment, to the larger environment, and to products produced. This is important because it is how safety and health become part of the organization's culture (comprised of expectations of how things are done, what is acceptable, how employees conduct themselves, and the like). The goal is to have safe and healthy work practices become an ingrained part of the organization's culture so that the right way becomes the expected way, and the expected way becomes the only acceptable way. When this happens, peer pressure and self-regulation will become the organization's most effective enforcers of safe and healthy work practices.

Not Concerned with Numbers

The vision statement is broad and general in nature. It is a philosophical statement, not a statement of specific quantifiable factors. Such statements as "improve the company's safety record by 50 percent" or "reduce lost-time-to-accidents by 150 hours" have no place in an organization's vision. These are objectives, which come later. The vision may contain qualitative terms (e.g., *best, worldclass, peak performance, excellent, high-quality*), but not quantitative terms (e.g., *15 percent, 130 hours, 15 accidents, $156,000*).

DEVELOPING THE SAFETY AND HEALTH VISION

It is possible for the TSM Facilitator to develop the organization's safety and health vision, but it is not advisable. There are three reasons why the *unilateral* approach is not recommended. First, having the TSM Facilitator work alone in developing the vision sends the wrong message to the members of the TSM Steering Committee. This approach says, "Safety is the job of the TSM Facilitator. You don't need to be involved." A vision developed unilaterally by the TSM Facilitator is likely to be viewed as his or her vision, not that of the organization. Second, this approach suggests that the TSM Facilitator, working alone, can develop a better vision than can the Steering Committee working as a team. Finally, the members of the TSM Steering Committee are more likely to buy into the vision if they are responsible for developing it. For these three reasons, the recommended approach for developing the safety and health vision is to have the TSM Steering Committee do it.

A question that is sometimes asked at this point is, "What about the organization's employees? Do they have any input into the development of the safety and health vision?" This is a good question because everything about TSM suggests the inclusive approach. However, development of the vision is an exception. Because of the commitments required, only the organization's executive managers have the authority necessary to approve the vision. Whether the TSM Steering Committee consists of the organization's executive managers or is a delegated committee, ultimate approval of the vision must come from the chief executive officer and his or her executive management team. Through the vision, the organization's executives express their commitment—relating to safety and health—to employees, customers, suppliers, and the community.

Although the TSM Facilitator should not develop the safety and health vision unilaterally, it is appropriate for this individual to present a *trial balloon* to get discussion started. In addition, it is the TSM Facilitator's job to ensure that all members of the Steering Committee understand the characteristics of a well-written vision, and that the final vision has these characteristics.

The vision statement itself has three components (Figure 5–2). The first component speaks to the organization's concern for its employees. The second component speaks to the organization's concern for customers. The third speaks to concern for the community at large. The example used earlier of the safety and health vision for The Douglas Corporation contains these three components:

■ *Employee component.* This component in the example reads as follows: *"provide an accident-free work environment that is conducive to productivity, quality, and competitiveness."* The accident-free work environment is provided for employees so

Figure 5–2
Components of a Safety and
Health Vision

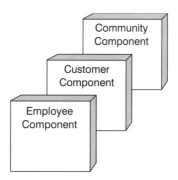

that they can focus all of their attention and energy on consistently achieving peak-performance levels.

■ *Customer component.* This segment in the example stresses the firm's commitment to manufacturing "*hazard-free products.*" The organization focuses on producing hazard-free products because it is concerned about the well-being of its customers.

■ *Community component.* This component in the example concentrates on Douglas's pledge to "*environmentally friendly processes.*" In producing its products, the organization attempts to make its processes as environmentally friendly as possible out of concern for the communities in which its facilities are located.

A vision that contains these three components and is easily understood; briefly stated; challenging, yet possible; stirring; tone setting; and unconcerned with numbers stands an excellent chance of being an effective vision.

GUIDING PRINCIPLES DEFINED

Guiding principles are values statements. They paint a word picture of an organization's corporate culture that guides the behavior of the organization as it pursues its vision. From the perspective of safety and health, guiding principles should describe clearly and succinctly an organization's values relating specifically to safety and health. The guiding

TSM TIP

A Safety and Health Example

The DuPont Corporation is setting an example of how firms can move responsibly to do their part in maintaining a safe and healthy environment. DuPont has adopted a corporate policy to ensure that all of its facilities operate in accordance with an identical set of safety, health, and environmental standards.[2]

principles contained in a safety and health plan should guide the behavior of all employees at all levels as they work together in pursuit of the organization's safety and health vision. Figure 5–3 contains the guiding principles for The Douglas Corporation.

PURPOSE OF THE GUIDING PRINCIPLES

A careful examination of these examples shows the purpose that guiding principles serve in an organization. The first four principles make clear to all employees that safety and health are high priorities. These simple statements can help resolve some of the most common situations that mitigate against safety and health in an organization. One such situation is the employee who, out of impatience, ignores or bypasses required safety procedures. Another situation is the supervisor who, under pressure to meet deadlines, encourages employees to ignore safety procedures. Yet another situation is the safe employee who is reluctant to challenge unsafe employees.

When there is a widely published guiding principle that says *"The safe way is the right way,"* these situations become less difficult to handle. The impatient employee knows that he has neither the stated support nor the tacit approval of the organization in taking dangerous shortcuts. The employee being pressured by her supervisor has

Figure 5–3
Sample Statement of Guiding
Principles

The Douglas Corporation
Guiding Principles for Safety and Health

1. At The Douglas Corporation, the *safe way* is the *right way.*
2. The Douglas Corporation is committed to safe and healthy processes and work practices.
3. At The Douglas Corporation, all employees at all levels are expected to do the following: (a) follow all applicable safety regulations; (b) take all precautions necessary to prevent injuries; and (c) correct other employees at any level who violate safety and health regulations.
4. At The Douglas Corporation, managers and supervisors are expected to enforce safety and health policies consistently and without exceptions.
5. At The Douglas Corporation, employees at all levels are expected to point out and help correct hazardous conditions.
6. The Douglas Corporation will not knowingly produce a hazardous product.
7. The Douglas Corporation will not knowingly harm the environment, the public, employees, or any other stakeholders.

something concrete to back her adherence to safety procedures, and the safe employee can simply point out the guiding principle to unsafe coworkers. Organizations get caught in pressure situations that, on the surface, seem to pit safety procedures against profits. This sometimes happens, for example, when sales and marketing personnel promise unrealistic delivery dates in order to close a deal. These circumstances often create pressure that, in turn, promotes the taking of shortcuts and other unsafe behavior. The guiding principles act as a shield against such short-sighted behavior.

When there is a widely publicized guiding principle that says *"Employees are expected to point out and help correct hazardous conditions,"* all employees at all levels know that safety is everybody's job. Such a guiding principle works against situations in which employees see safety as someone else's job, or who respond to unsafe conditions by saying, "That's not my job."

A guiding principle that says, *"The Douglas Corporation will not knowingly produce a hazardous product,"* sets the tone for customers, employees, and suppliers. Customers know that if some aspect of a product produced by The Douglas Corporation turns out to be hazardous, they won't be stonewalled or given the cold shoulder when they point out the problem. Employees know that if a potentially hazardous condition is discovered during production, their responsibility is to point it out and help correct it. Suppliers know that the products they provide to The Douglas Corporation must adhere to the company's guiding principles. Providing faulty or unsafe material to The Douglas Corporation clearly will not be tolerated.

A guiding principle that holds that the Corporation *"will not knowingly harm the environment"* sets the tone for the organization's community relationships. The Douglas Corporation's neighbors can expect the company to be friendly to the environment, and they can expect an open door if concerns arise.

DEVELOPING THE GUIDING PRINCIPLES

Guiding principles describe what is acceptable behavior as the organization pursues its vision. By writing down their guiding principles, organizations remove any doubt that they are committed to safety and health. Because they represent organizational commitment, the guiding principles are developed by the TSM Steering Committee. No one but executive level managers can commit an organization to the type of values described in a set of guiding principles.

=== TSM CASE STUDY ===

What Would You Do?

As Vice President for Production, Marvin Rosemont serves on the TSM Steering Committee for Western Prestressed Concrete (WPC). The department in WPC that is most likely to be affected by injuries to employees is the Production Department. Consequently, Rosemont is determined to see the TSM Steering Committee develop a clear safety and health vision and a comprehensive set of guiding principles. However, his colleagues on the Steering Committee don't seem to share his enthusiasm. They understand that a vision and guiding principles are needed. But they want to delegate the project to WPC's TSM Facilitator. If you were in Rosemont's position, what would you tell your colleagues to get them more involved?

In developing guiding principles, it is important for Steering Committee members to understand that the organization's concern for safety and health must encompass not just employees, but all stakeholders including customers, suppliers, contractors, and the community. A well-written set of guiding principles will indicate that: (a) the organization is committed to safety for employees, customers, and the community; and (b) that all stakeholders including employees, suppliers, contractors, customers, and the community are expected to play their respective parts in helping the organization establish and maintain a safe and healthy work environment.

STEP 5 IN ACTION

Mack Parmentier has mixed emotions about developing the safety and health vision and guiding principles for MPC. So far the process has been both fascinating and frustrating. On the one hand, the TSM Steering Committee is making progress—finally. But before they could begin developing a safety and health vision with accompanying guiding principles, the Steering Committee had had to develop a strategic plan for MPC. The company had been operating for years without any strategic direction from the top.

Parmentier had experienced some difficulty getting several committee members to understand that he has no role to play in developing the company's strategic plan. This is a job for the company's executive managers. He has also experienced some difficulty in helping Steering Committee members understand that the safety and health vision/ guiding principles could not be developed until the broader corporate vision, mission, guiding principles, and objectives were in place.

Waiting for MPC's executives to develop a strategic plan for the company had set Parmentier's implementation schedule back at least two weeks, but things are moving in the right direction now. MPC now has a strategic plan, and the Steering Committee is in the process of developing a safety and health vision and guiding principles. Parmentier hopes that having gone through both processes, the TSM Steering Committee may actually end up understanding the relationship between the corporate vision and the safety/ health vision even better than executives in companies that begin their TSM implementation with a strategic plan in place.

SUMMARY

1. The safety and health vision describes the organization's dream concerning safety and health. It represents a beacon in the distance toward which the organization is always moving.

2. The safety and health vision serves as the guiding force in an organization for ensuring a safe and healthy work environment. It shows not just that management is committed, but what specifically management is committed to.

3. A well-written safety and health vision will have the following characteristics: it will be easily understood, briefly stated yet clear and comprehensive, challenging yet possible, stirring, tone setting, and unconcerned with numbers.

4. The safety and health vision should be developed by the TSM Steering Committee. Asking the TSM Facilitator to develop the vision sends the wrong message to employees. Asking employees to write the vision is inappropriate because only executive managers can make the commitments implicit in the vision.

5. A safety and health vision has three components: employee, customer, and community components.

6. Guiding principles are values statements that paint a word picture of an organization's corporate culture. From the perspective of safety and health, guiding principles should describe clearly and succinctly an organization's values relating specifically to health and safety.

7. The organization's guiding principles for safety and health describe what is acceptable behavior as the organization pursues its vision. Because they represent organizational commitment, the guiding principles are developed by the TSM Steering Committee.

KEY TERMS AND CONCEPTS

Briefly stated	Guiding principles
Challenging	Not concerned with numbers
Community component	Safety and health vision
Customer component	Sets the tone
Easily understood	Stakeholder
Employee component	Stirring

REVIEW QUESTIONS

1. Explain the concept of a safety and health vision.
2. What is the purpose of the safety and health vision?
3. Write a sample safety and health vision for a hypothetical organization.
4. List the characteristics of a well-written safety and health vision.
5. Describe how the safety and health vision should be developed.
6. Explain the three components of a safety and health vision.
7. What is a guiding principle?

8. What is the purpose of guiding principles?

9. Describe how guiding principles are developed.

ENDNOTES

1. David L. Goetsch and Stanley B. Davis, *Implementing Total Quality* (Upper Saddle River, N.J.: Prentice Hall, 1995), p. 106.

2. David L. Goetsch, *Occupational Safety and Health in the Age of High Technology*, 2nd ed., (Upper Saddle River, N.J.: Prentice Hall, 1996), p. 29.

Develop the Organization's Safety and Health Mission and Objectives

- The Mission: Definition and Purpose
- Developing the Mission
- Rationale for Broad Objectives
- Nature of Broad Objectives
- Establishing Broad Objectives
- Step 6 in Action

In Step 5 the TSM Steering Committee created the organization's safety and health vision and guiding principles. Step 6 takes the Steering Committee through the next sequential activity in the planning phase of the TSM implementation: developing a mission and a set of broad objectives for safety and health. Although the term *broad objectives* is used here, the term *goals* is interchangeable and could also be used in this context.

THE MISSION: DEFINITION AND PURPOSE

The vision developed in the previous step described the organization's safety and health dream. A vision is just that, a dream. A mission, on the other hand, is an entity's reason for existing. It describes the entity's purpose and responsibilities. The entity in this case is the TSM Steering Committee. In other words, whereas the vision describes the committee's dream for an ideal work environment, the mission describes the committee's purpose and responsibilities in making the dream a reality. It answers the following questions about the TSM Steering Committee: Who are we? What do we do? What are our responsibilities? Figure 6–1 illustrates the concept.

The previous step used The Douglas Corporation to illustrate the concepts of the vision and guiding principles. The present step, and all successive steps, will use this

Figure 6–1
Writing the Mission Statement

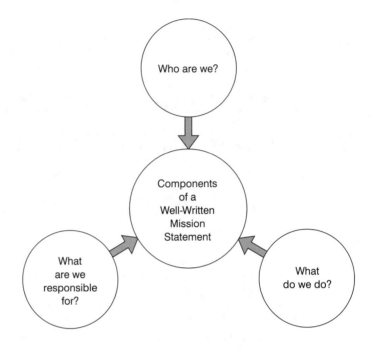

same organization for illustrative purposes. Figure 6–2 contains the mission statement for The Douglas Corporation's TSM Steering Committee. In this example one can find all three of the components of a well-written mission statement.

The first sentence in Douglas's mission statement reads, "The TSM Steering Committee consists of The Douglas Corporation's executive-level decision makers." This is the *who-are-we* component. The *what-do-we-do* component is found in the second sentence, which reads "The mission of the TSM Steering Committee is to ensure that the company's safety and health vision is accomplished, the company's safety and health guiding principles are adhered to, and the company's safety and health objectives are accomplished." The *what-we-are-responsible-for* component is spelled out in the following list of specific duties:

- Building safety and health concerns into the company's strategic plan
- Developing safety and health policies for the company
- Overseeing the company's safety and health program
- Approval of recommendations for improvements to the work environment
- Allocation of resources for safety and health
- Assignment of charters for improvement project teams
- Making safety and health part of the company's performance appraisal and reward/ recognition processes

**The Douglas Corporation
TSM Steering Committee
Safety and Health Mission Statement**

The TSM Steering Committee consists of The Douglas Corporation's executive-level decision makers. The purpose of the TSM Steering Committee is to ensure that the company's safety and health vision is accomplished, the company's safety and health guiding principles are adhered to, and the company's safety and health objectives are achieved. Specific responsibilities include the following:

- Building safety and health concerns into the company's strategic plan
- Developing safety and health policies for the company
- Overseeing the company's safety and health program
- Approval of recommendations for improvements to the work environment
- Allocation of resources for safety and health
- Assignment of charters for improvement project teams (IPTs)
- Making safety and health part of the company's performance appraisal and reward/recognition processes

Figure 6–2
Sample TSM Mission Statement

DEVELOPING THE MISSION

The safety and health mission is developed by the TSM Steering Committee. The mission in question is that of the Steering Committee. As with the vision, the mission is a statement of commitment that can be made only by an organization's executive-level managers. A well-written mission statement has three broad components (see Figure 6–1). It is important that all three be present in the mission statement regardless of the entity in question.

TSM TIP

Job of the Safety and Health Manager

The job of the modern safety and health manager includes the following types of tasks: facilitation, analysis, prevention, planning, evaluation, promotion, and compliance.

Who-Are-We Component

This component describes the entity to which the mission applies. This is especially important in the case of the TSM Steering Committee. It lets all stakeholders know the composition of the committee, which is—in itself—a powerful indicator of the organization's commitment to safety and health. This is yet another reason why the author recommends that the TSM Steering Committee be comprised of the organization's executive-level managers.

In developing the mission statement, members of the TSM Steering Committee should endeavor to keep the *who-are-we* component concise and to the point. Typically one sentence is sufficient. What follows are examples of this component from the safety and health mission statements of several different TSM Steering Committees:

■ The TSM Steering Committee is the company's executive management team augmented by the Directors of Safety/Health and Human Resources.

■ The Executive Council of Jones Products, Inc. doubles as the TSM Steering Committee.

■ Rayburn Company's CEO and corporate vice presidents serve as the TSM Steering Committee.

What-We-Do Component

This component describes the purpose of the entity in question. It has at least three elements, as depicted by Figure 6–3. The *what-we-do* component of the mission statement for every TSM Steering Committee must include these three elements. This is because no matter what the type of organization, its TSM Steering Committee's purpose is threefold and the same: (a) ensure that the organization's safety and health vision is accomplished; (b) ensure that the organization's guiding principles for safety and health are

Figure 6–3
Three Elements of the "What-We-Do" Component

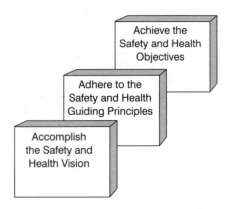

TSM TIP

Supervisors and Safety

Supervisors play a key role in the maintenance of a safe and healthy workplace. Consequently, safety and health professionals should ensure that all supervisors have the training they need to be positive agents of workplace safety.

---

adhered to; and (c) ensure that the organization's safety and health objectives are achieved. These three elements don't change. There can be additional elements, but these three must be present.

What-We-Are-Responsible-For Component

This component describes in specific terms what the TSM Steering Committee is responsible for doing. This is the most specific component in the mission statement. Although it can vary somewhat from organization to organization, the specific responsibilities listed in Figure 6–2 are typical of those assigned to TSM Steering Committees. These responsibilities relate to strategic planning, policy development, oversight, approval of recommendations, resource allocation, assignment of charters, performance appraisal, and reward/recognition processes.

RATIONALE FOR BROAD OBJECTIVES

Broad objectives take the next logical step beyond the mission and answer the question, "What does the TSM Steering Committee plan to achieve?"

The broad objectives translate the TSM Steering Committee's mission into more specific terms that can be monitored and measured. They represent the Steering Committee's best thinking concerning what must be done in order to accomplish its mission. Consequently, there should be one or more objectives for each of the three elements of the mission's *what-we-do*. This component describes in broad and general terms what the TSM Steering Committee is chartered to do, i.e., its reason for being. The objectives relating to this component of the mission translate the component into more specific measurable terms.

================= TSM CASE STUDY =================

What Would You Do?

Mark Underwood had been caught off guard at Belco Corporation's last TSM Steering Committee meeting. Although Underwood has attended several seminars on TSM at safety and health conferences, he has no practical experience in facilitating a TSM implementation. Belco's Steering Committee had convened to develop a mission statement and a set of broad objectives. The meeting had barely gotten underway when two participants got into a disagreement over the differences between the vision and the mission. As the TSM Facilitator, Underwood had been called upon to settle the question. If you found yourself in this situation, how would you explain the differences?

NATURE OF BROAD OBJECTIVES

Figure 6–4 is a continuum of the various components of the TSM Steering Committee's plan for safety and health. The farther one goes to the left on this continuum, the more *what-oriented* the component becomes. Consequently, the vision component of a TSM Steering Committee's safety and health plan is 100 percent *what* in its orientation. No mention is made in the vision of how it will be accomplished. The mission is also *what* oriented, but because it is to the right of the vision on the continuum it begins to take on shades of *how*. On the one hand the mission describes *what* the TSM Steering Committee is chartered to do. But on the other hand, if one asks, "How will the Steering Committee achieve its vision?" the best answer is, "By efficiently and effectively accomplishing its mission." Therefore, the mission is a *what-oriented* statement with *how* connotations relating to the vision.

This same relationship exists between the mission and the broad objectives. The broad objectives are *what-oriented* statements that show *how* the mission will be accomplished. One deals 100 percent in *how-to* terms only at the level of specific strategies that are typically developed at the department level. Strategies are written in such specific terms that their accomplishment can actually be observed or, at least, easily quantified and measured.

Figure 6–5 summarizes the characteristics of broad objectives. Broad objectives should flow directly out of the mission and support its accomplishment. If a broad objective cannot be tied directly to the mission, one of two things should happen. Either the objective should be modified or it should be dropped altogether. If an objective does not tie directly to the mission, but it is judged to be valid, the mission should be revised.

Figure 6–4
Planning Continuum

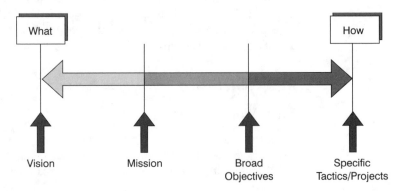

Figure 6–5
Characteristics of Broad
Objectives

Broad objectives of an organization always:

- Flow directly out of the mission.
- Support accomplishment of the mission.
- Apply to the entire organization, rather than to individual departments or subunits.
- Are nonrestrictive in nature.
- Describe *what* is to be accomplished.
- Remain relatively static.

Broad objectives apply to the entire organization rather than to individual departments or subunits. Department strategies for accomplishing the broad objectives are the final component on the planning continuum (Figure 6–4). Consequently, broad objectives—like the vision, guiding principles, and mission—are developed by the TSM Steering Committee.

Broad objectives are enabling in nature, not restrictive. They should never hold an organization back. This means they must be open-ended in terms of what is to be accomplished, as in the following example:

We will produce only hazard-free products.

This broad objective is open-ended in that it contains no restrictions. It doesn't say "most of our products will be hazard-free," or "50 percent of our products will be hazard-free." Numbers may be used in writing broad objectives, but only if used with care. They have a way of minimizing rather than maximizing performance. There is a human tendency to say "That's good enough," once the numerical value in an objective has been reached.

Broad objectives describe what is to be accomplished by the entity in question, in this case, the TSM Steering committee. They have somewhat of a how-to orientation because they answer the question, "How will we accomplish our mission?" The answer to this question should be, "By achieving the following objectives." Nonetheless, broad objectives are stated in terms of what is to be accomplished.

TSM TIP

Mechanical Injuries on the Job

The most common mechanical injuries on the job involve cutting, tearing, shearing, crushing, breaking, straining, spraining, and puncturing.

Broad objectives are relatively static. This means, like the vision and mission, they don't change much over time. Since they are written in an open-ended format, broad objectives represent what the entity in question is always striving to reach but seldom does, or at least does not reach fully. In addition, even if a broad objective is accomplished, it is likely to remain a valid objective. For example, if an organization sets a broad objective of eliminating back injuries, and then does so in a given year, it will still retain the objective. Otherwise, back injuries will begin to reoccur.

It is the strategies developed by IPTs that change with relative frequency. Such strategies are set at the team level, implemented, and monitored. If they achieve the desired result, they may be retained or replaced by other strategies. If they don't, they are modified or discarded altogether. Team-level strategies are not dealt with in this discussion because they are not developed by the TSM Steering Committee and because they vary so widely from organization to organization, depending on individual circumstances.

Figure 6–6 contains the broad objectives established by the TSM Steering Committee of The Douglas Corporation. These objectives flow directly out of that portion of the company's mission statement that reads as follows: "The purpose of the TSM Steering Committee is to ensure that the company's safety and health vision is accomplished." The broad

The Douglas Corporation
TSM Steering Committee
Safety and Health Broad Objectives

- Establish a work environment that is free of ergonomic hazards.
- Establish a work environment that is free of fall/acceleration-related hazards.
- Establish a work environment that is free of lifting hazards.
- Establish a work environment that is free of heat/temperature hazards.
- Establish a work environment that is free of pressure hazards.
- Establish a work environment that is free of electrical hazards.
- Establish a work environment that is free of fire hazards.
- Establish a work environment that is free of toxic substance hazards.
- Establish a work environment that is free of explosives-related hazards.
- Establish a work environment that is free of radiation hazards.
- Establish a work environment that is free of noise hazards.
- Establish a work environment that is free of vibration hazards.
- Establish a work environment that is free of automation-related hazards.
- Establish a work environment that is free of bloodborne hazards.

Figure 6–6
Sample of a Corporation's Broad Objectives for TSM

objectives support the accomplishment of The Douglas Corporation's mission, apply to the entire organization, and are nonrestrictive in nature. They describe what the TSM Steering Committee of The Douglas Corporation plans to accomplish, and they are relatively static.

ESTABLISHING BROAD OBJECTIVES

The TSM Steering Committee, working with the TSM Facilitator, develops the broad objectives. A key role played by the facilitator is ensuring that all applicable safety and health issues are properly covered by one or more objectives. For example, if the organization—because of its processes—is concerned about mechanical, falling, acceleration, and lifting hazards, the TSM Facilitator should ensure that all of these areas are spoken to in the objectives.

Figure 6–7 is a model that steering committees can use when developing broad objectives. In this model's first step, all suggestions are treated as being valid. Although

Figure 6–7
Model for Developing Broad
Objectives

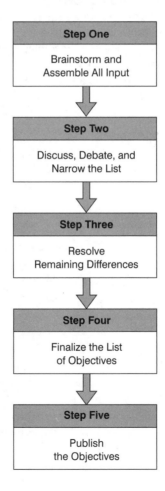

Step One

Brainstorm and
Assemble All Input

Step Two

Discuss, Debate, and
Narrow the List

Step Three

Resolve
Remaining Differences

Step Four

Finalize the List
of Objectives

Step Five

Publish
the Objectives

questions may be asked for clarification, neither discussion nor judgmental comments are allowed at this point. The facilitator simply records and displays—on a flip chart or marker board—all input from the group.

In the second step, participants discuss the individual suggestions, one at a time, debating the merits of each. Suggestions that cannot withstand the scrutiny of the group are discarded, and the list of potential objectives is narrowed. In the third step, the list of potential objectives is divided into two categories: those that have a consensus and those that still require further discussion. The facilitator places a check mark or some other visual indicator next to the consensus objectives. All others are discussed at length until remaining differences among participants have been resolved.

In the fourth step the facilitator distributes a list of cautions to be observed when developing broad objectives and reviews all entries on the list with participants. Figure 6–8 is a graphic listing some relevant concerns. The cautions illustrated are used by the Steering Committee to convert the draft objectives that have achieved consensus into final form. Once this has been accomplished, the objectives are ready to be published as part of the organization's safety and health plan.

The cautions set forth in Figure 6–8 are important. They should be observed carefully when developing broad objectives. To begin, keep the broad objectives relatively few in number. A good rule of thumb to follow is, "If you have more than fifteen broad objectives, check to see if any of them are actually strategies that should be dealt with at the

Figure 6–8
Cautions to Observe When
Writing Broad Objectives

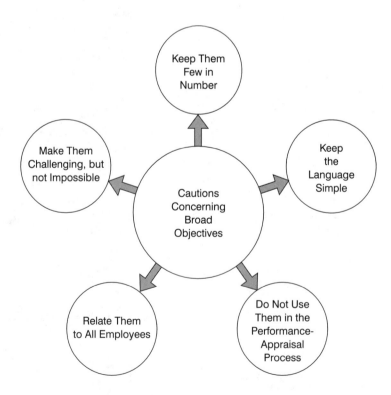

TSM TIP

Workers' Compensation Laws

All fifty states in the United States have workers' compensation laws, but the laws vary markedly. The purpose of workers' compensation laws is to provide benefits, pay medical costs, provide rehabilitation, decrease litigation, and encourage safety. Whether or not the laws have accomplished these goals is a controversial question.

IPT level." This is not a hard and fast rule. Some organizations may be so diverse with regard to processes that they need more than fifteen broad objectives to cover all applicable safety concerns. But such situations are rare. Generally, a number higher than fifteen should raise a yellow flag of caution.

Keep the language of broad objectives simple enough that they can be easily understood by all stakeholders. Avoid organization-specific language that outsiders such as suppliers might not understand. Stakeholders can help the organization accomplish its objectives only if they understand the objectives.

Broad objectives should not be used in employee performance appraisals. Individual department or subunit strategies may be, but not broad objectives. This is because properly written objectives are too broad to hold any individual or even department responsible for accomplishing them. Remember, broad objectives apply to the entire organization.

Broad objectives should be written so that they relate to all employees rather than to a single group, department, or subunit. An objective that is so specific that it relates to a single group is probably a strategy, not a broad objective. Such an objective should either be rewritten in broader terms, or reworked for use as a strategy.

Broad objectives should on the one hand challenge an organization, but on the other, not be impossible to achieve. The organization and its employees should have to stretch in order to accomplish the broad objectives. However, people and organizations are like rubber bands in that they can stretch only so far. If the broad objectives established for an organization are seen by employees as being impossible, they will simply ignore them.

BROAD OBJECTIVES AS A BRIDGE

In a TSM setting, executive managers serving as the TSM Steering Committee develop the strategic aspects of the organization's safety and health plan (i.e., its vision, guiding principles, mission, and broad objectives). The TSM Facilitator then works with individual IPTs to develop the tactical aspects of the plan. In other words, the specific tactics for accomplishing the broad objectives are developed by IPTs working in conjunction with the TSM Facilitator. The bridge between the strategic level of planning (what) and the tactical level (how) consists of the broad objectives, as shown in Figure 6–9.

Figure 6–9
Bridging the Gulf between
Strategy and Tactics

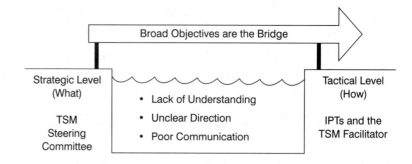

IPTs are chartered by the TSM Steering Committee to undertake specific projects that will move the organization closer to accomplishing its broad objectives. For example, one of the Douglas Corporation's broad objectives (from Figure 6–6) reads as follows:

Establish a work environment that is free of mechanical hazards.

An IPT might be chartered to identify mechanical hazards in the organization's precision machining department, and make recommendations for eliminating those hazards. Another IPT might be chartered to undertake a similar project—simultaneously—in another department. Such teams and projects are ongoing as the organization seeks to improve the work environment continually.

=========== STEP 6 IN ACTION ===========

"I don't understand how it all fits together," said Martha Scott, MPC's Director of Human Resources. Several other participants nodded their heads in agreement. "We have a safety and health vision and guiding principles. We have a mission for the TSM Steering Committee and broad objectives. But how do we put it all to work? It seems to me that the next step should be to have all departments develop a plan of work for carrying out some piece of our larger plan," said Scott.

"That is very close to what will eventually happen, Martha," was Mack Parmentier's response. "But, first, there are some other steps we need to take." Parmentier then went on to explain that eventually all employees would be fully informed concerning the safety and health plan developed by the TSM Steering Committee. IPTs would be formed and chartered by the Steering Committee to do what Martha Scott had proposed the departments do. He explained that these IPTs will be both natural work teams within departments and cross-functional teams with members from different departments.

He explained that once IPTs began to be chartered, there would always be teams working on various safety and health concerns that are tied directly to the Steering Committee's plan, and that the Steering Committee would monitor the work of these teams. But before getting to that point, there is a good deal more work to be done, and if any step in the process is skipped, the whole process might fail.

Parmentier closed his portion of the meeting by saying, "I know you are all anxious to get started on improving the company's safety record as well as other issues that are hurting our performance. Don't worry, we will begin to initiate hard and fast improvements soon. But let's don't skip any steps. We've come a long way. Stay the course, and before long you'll have the pleasure of watching this company transform itself into a worldclass competitor. Our next step is to get the word out about the safety and health plan. We need to share the plan with every employee and make sure they understand it. With your permission, I'll get the ball rolling right now."

SUMMARY

1. The safety and health mission describes the TSM Steering Committee's reason for existing, its purpose, and its responsibilities. It should answer the following questions about the TSM Steering Committee: Who are we? What do we do? What are our responsibilities?

2. The broad objectives for safety and health represent the TSM Steering Committee's best thinking concerning what must be done in order to accomplish the safety and health mission. They translate the broad objectives into more specific terms.

3. Broad objectives describe what the organization plans to accomplish, not how it plans to do so. However, the objectives do take on shades of how-to in the sense that they describe how the safety and health mission will be accomplished.

4. Broad objectives should flow directly out of the mission, apply to the entire organization, be enabling in nature rather than restrictive, describe what is to be accomplished, and be relatively static in nature.

5. When developing broad objectives it is a good idea to hold the number developed to fifteen or fewer. An organization that requires more than fifteen may be developing specific strategies rather than broad objectives. This rule of thumb is not hard and fast. There may be circumstances in which more than that number of broad objectives is acceptable. However, a large number of broad objectives should always at the very least raise a caution flag.

6. Broad objectives serve as a bridge between the work of the TSM Steering Committee and the work that will be done by the IPTs. The projects taken on by IPTs are for the sole purpose of moving the organization toward accomplishment of its broad objectives.

KEY TERMS AND CONCEPTS

Apply to the entire organization	Keep the language simple
Broad objectives	Mission
Flow directly out of the mission	Open-ended

Relatively static   What-we-do

Vision   Who-we-are

What-we-are-responsible-for

======================= REVIEW QUESTIONS =======================

1. Explain the purpose of the safety and health mission.
2. What is the difference between the vision and the mission?
3. Explain the three components of the safety and health mission.
4. Are broad objectives *what* or *how* oriented? Explain.
5. List and explain the cautions to remember when developing broad objectives for safety and health.
6. What is meant by the following statement? Broad objectives serve as a bridge between the TSM Steering Committee's work and that of the IPTs.
7. Describe the three elements of the *what-we-do* component of the safety and health mission.

Communicate and Inform

In Step 6, the TSM Steering Committee completed the safety and health plan by developing the mission and broad objectives. The complete safety and health plan includes the vision, guiding principles, mission, and broad objectives. The present step, Step 7, calls for informing all stakeholders of the plan. It is the last step in the *planning and preparation* phase of the TSM implementation process.

RATIONALE FOR COMMUNICATING AND INFORMING

Employees are people, and people like to be informed. One of the fastest ways to damage morale in an organization is to fail to keep employees fully informed concerning issues and decisions that affect them. The planning activities that have been taking place during the initial phase of the TSM implementation process will have been noticed by employees. It will have been noticed that executive managers are meeting more frequently than usual, and that some of the meetings have taken place off site.

Employees are keenly interested in the comings and goings of executive managers. In particular, they will wonder, "What are our executives doing and how will it affect me?" As soon as employees start asking themselves and each other such questions, the rumor mill will move into high gear. Rumors are a fact of life anytime people interact in a group setting. They are a product of human nature, a fact managers must understand and accept. However, accepting their inevitability does not mean that managers should accept the problems rumors can cause. Figure 7–1 shows five problems that rumors can cause in the workplace.

Stress is a natural human reaction to a threatening situation. One of the major causes of stress in the workplace is job insecurity. In this age of downsizing, re-engineering, and corporate restructuring, job insecurity is a common phenomenon—one that is likely to be intensified if rumors begin to circulate about an unusual number of unexplained executive-level meetings. Stress that is left unrelieved can damage the morale of employees and lead to panic reactions as feelings of job insecurity get blown out of proportion. This, in turn, can lead to a reaction that is becoming increasingly common and increasingly dangerous. This reaction is anger and its cohort, hostility. Taken together, these problems typically lead to a higher than normal level of turnover, and can have even worse consequences.

Preventing the types of problems just described is part of the rationale for communicating and informing. There is actually a twofold rationale. First, communication is the most effective way to shut down the rumor mill and prevent the problems it can cause. Second, since employees are key stakeholders in the organization, they have a major role to play in carrying out the safety and health plan. They can play this role only if they are fully informed concerning all aspects of the plan.

WHY WAIT TO COMMUNICATE?

If the rumor mill can cause such serious problems, why wait until Step 7 to communicate with employees about the safety and health plan? The principle reason is that it would be premature to do otherwise. Until Step 6 is completed there is no plan to inform employees about. Trying to develop a plan and simultaneously keep employees informed

Figure 7–1
Problems That Can Be Caused
by Rumors

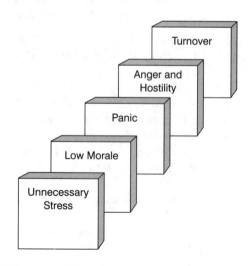

TSM TIP

Ergonomic Hazards in the Workplace

"The proliferation of uncomfortable and dangerous conditions in the workplace creates ergonomic hazards. Assessing ergonomic hazards involves checking for job designs that require unnatural positions or posture, excessive wasted motion, or frequent manual lifting."[1]

---

about it would lead to confusion and a proliferation of misinformation. Such an approach would probably cause more harm than good.

This does not mean, however, that executives should carry out their planning activities under a shroud of mystery or secrecy. A better approach is to let employees know early on that executive managers are developing a safety and health plan that will be distributed in the near future. There are a number of ways to get this message across, including but not limited to the following: a memorandum to all employees from the CEO, an announcement in a company publication such as a newsletter, verbal communication within departments initiated by each respective member of the TSM Steering Committee, bulletin board announcements, and/or employee meetings.

The method or methods chosen will depend on the size and complexity of the organization and the preferences of executive managers. In any case, it is a good idea to use more than one method. Generally speaking, the more communication methods used, the more employees who will receive the message and receive it accurately.

WHO TO TELL AND HOW

An organization cannot communicate too much. Even in organizations that pride themselves on communicating, many employees will claim "There is too little communication" when responding to employee-satisfaction surveys. Organizational communication is a difficult undertaking that is fraught with roadblocks. Some employees will invariably ignore or throw away memorandums from the CEO, viewing these attempts at commu-

---

TSM TIP

Safety, Health, and Peak Performance

"Get more than one opinion before you announce a decision. Let everyone who will be affected by the resultant change in process or procedure know that you are considering a change. Solicit input."[2]

---

nication as *junk mail.* Others won't read company newsletters or bulletin board notices. Still others will miss meetings in which announcements are made. It seems that no matter how hard managers try, there will always be employees who don't get the word. This is why it is so important to use more than one communication medium when trying to convey an important message.

Once the safety and health plan has been completed, it should be communicated to all employees using a variety of different methods. An outline of the substance of such a communication is shown in Figure 7–2. In addition to informing employees of the safety and health vision, guiding principles, mission, and broad objectives, the CEO or his/her designees should explain why the plan was developed and what it means to employees.

Figure 7–3 is the safety and health plan The Douglas Corporation (TDC) shared with all of its employees. In addition to sharing the plan itself, it is important to provide an explanation of why the plan was developed and what it means to employees. Figure 7–4 contains the explanations provided to all employees of The Douglas Corporation. The explanations were attached to the safety and health plan that was distributed to IPT members. In addition to distributing the plans and explanations, TDC's executives met with the employees in their respective departments to answer questions.

Figure 7–2
Outline of Subjects to Be
Communicated

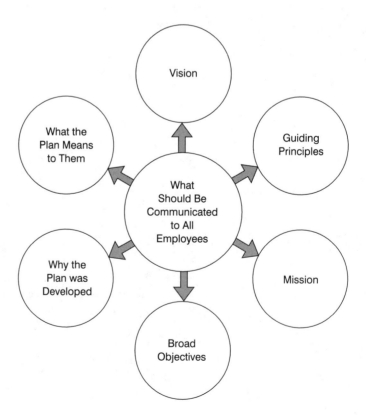

TSM TIP

Corporate Image and the Environment

"Image problems are not limited to product liability issues. Companies that are not careful about protecting the environment may find themselves the subject of protest demonstrations on the nightly news. Such negative publicity can harm the corporate image and translate very quickly into market losses. As corporate decision makers become more sensitive to these facts, health and safety professionals can use them to gain a commitment to their programs."[3]

The TDC approach is an excellent example of how an organization can inform its employees of the safety and health plan, why it was developed, and what the plan means to them. Notice in Figure 7–4 that TDC executives based the rationale for their plan on the need to meet the challenge of global competitiveness. They also make it clear that a safe and healthy work environment is a key component in the company's ability to continually improve its products, processes, and services.

In explaining what the plan means to employees, TDC's executives inform them of the need to serve on IPTs. This alerts employees to the third phase of the implementation process and the critical role they will play in it. During face-to-face meetings with employees, members of the TSM Steering Committee can explain the IPT concept in greater depth and answer any questions that might come up.

COMMUNICATION METHODS

In the previous section, the point was made that more than one communication medium should be used when informing employees about the safety and health plan. Figure 7–5 shows several vehicles that can be effective for communicating with employees; this is especially so when two or more are used in concert with each other. Each of these methods has its own inherent advantages and disadvantages.

Verbal Presentations

Verbal presentations made by individual members of the TSM Steering Committee should be viewed as mandatory. Regardless of the other methods chosen, the communication package should include verbal presentations. Advantages of this method are as follows:

- Allows executives to apply the personal touch.
- Reinforces the speaker's commitment to the plan specifically and to safety and health in general.
- Allows the speaker to answer questions employees may have.

	The Douglas Corporation
TDC	**Corporate Vision, Guiding Principles for Safety and Health, Mission Statement, and Broad Objectives**

The Douglas Corporation will provide an accident-free work environment that is conducive to productivity, quality, and competitiveness; hazard-free products; and environmentally friendly processes.

Guiding Principles for Safety and Health

1. At The Douglas Corporation, the *safe* way is the *right* way.
2. The Douglas Corporation is committed to safe and healthy processes and work practices.
3. At The Douglas Corporation, all employees at all levels are expected to do the following: (a) follow all applicable safety regulations; (b) take all precautions necessary to prevent injuries; and (c) correct other employees at any level who violate safety and health regulations.
4. At the Douglas Corporation, managers and supervisors are expected to enforce safety and health policies consistently and without exceptions.
5. At the Douglas Corporation, employees at all levels are expected to point out and help correct hazardous conditions.
6. The Douglas Corporation will not knowingly produce a hazardous product.
7. The Douglas Corporation will not knowingly harm the environment, the public, employees, or any other stakeholder.

Mission Statement

The TSM Steering Committee consists of The Douglas Corporation's executive-level decision makers. The purpose of the TSM Steering Committee is to ensure that the company's safety and health vision is accomplished, the company's safety and health guiding principles are adhered to, and the company's safety and health objectives are achieved. Specific responsibilities include the following:

- Building safety and health concerns into the company's strategic plan
- Developing safety and health policies for the company

Figure 7–3
Sample Safety and Health Plan

- Overseeing the company's safety and health program
- Approval of recommendations for improvements to the work environment
- Allocations of resources for safety and health
- Assignment of charters for improvement project teams
- Making safety and health part of the company's performance appraisal and reward/recognition processes

Broad Objectives

The TSM Steering Committee of The Douglas Corporation is committed to accomplishing the following broad objectives:

- Establish a work environment that is free of ergonomic hazards.
- Establish a work environment that is free of fall/acceleration-related hazards.
- Establish a work environment that is free of lifting hazards.
- Establish a work environment that is free of heat/temperature hazards.
- Establish a work environment that is free of pressure hazards.
- Establish a work environment that is free of electrical hazards.
- Establish a work environment that is free of fire hazards.
- Establish a work environment that is free of toxic substance hazards.
- Establish a work environment that is free of explosives-related hazards.
- Establish a work environment that is free of radiation hazards.
- Establish a work environment that is free of noise hazards.
- Establish a work environment that is free of vibration hazards.
- Establish a work environment that is free of automation-related hazards.
- Establish a work environment that is free of bloodborne hazards.

Why the Safety and Health Plan was Developed

After many years of competing successfully with companies in the United States, The Douglas Corporation (TDC) now finds itself competing against companies from Europe, Canada, and Asia. This *globalization* phenomenon has markedly increased the level of competition we face on a daily basis. In order to continue to succeed and prosper, TDC will need every competitive edge it can manage.

The executive management team of TDC is in full agreement that a safe and healthy workplace will give our company a sustainable competitive advantage. Such an environment will allow employees to focus their full attention and all of their creative energy on continually improving our processes, products, and services. It will also eliminate situations in which our profits are drained off by the non-value added expenses associated with accidents and incidents. Our safety and health plan was developed to help TDC achieve a safe and healthy work environment that is conducive to peak performance and the continual improvement of productivity, quality, and competitiveness.

What the Plan Means to Employees

This plan means that all employees of TDC will have a safe and healthy workplace that is free of hazards and other factors that inhibit peak performance and continual improvement. In the near future, you may be asked to serve on *Improvement Project Teams* (IPTs) that are chartered to make recommendations for safety and health-related improvements in the workplace. With your help, we plan to ensure that TDC is a safe, healthy, competitive company.

Figure 7–4
Explanation Offered of Sample Plan

- Can provide feedback on concerns the TSM Steering Committee may have overlooked.

Although the verbal presentation should be considered mandatory, it does have drawbacks. These disadvantages are as follows:

- Can become burdensome to presenters if the presentation has to be repeated several times in order to reach all employees.
- Weak executives can let employees turn the presentation into a referendum on the safety and health plan.

VIDEOTAPED PRESENTATION

A videotaped presentation made by the organization's CEO can be an effective component of the overall communication package. Advantages of this method include the following:

Figure 7–5
Communication Methods

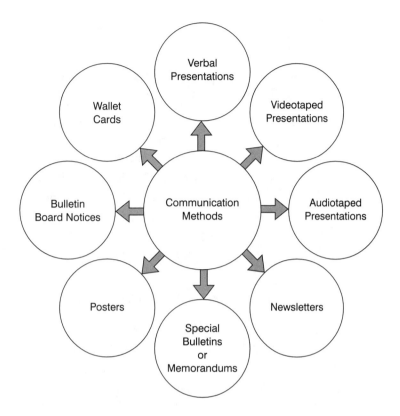

- Can convey the personal touch, if properly produced, although not as well as a verbal presentation.
- Can be retaped and edited until the message and the tone are just right.
- Can be repeated as often as necessary without burdening the speaker.
- Can be played at the convenience of the audience.
- Can be used as a substitute for an executive who might not handle the verbal presentation well.

Disadvantages of the videotaped presentation are as follows:

- Not as personal as a verbal presentation.
- Less likely than a verbal presentation to stir a response.
- Audience participation is impossible. Consequently, the benefits of questions, comments, concerns, and employee input are lost.
- Requires a video production capability.

Audiotaped Presentation

An audiotaped presentation made by the organization's CEO, although limited in its utility, can be part of the overall communication package. Some advantages of this method are that it:

- Can convey a small measure of the personal touch, although not as well as a verbal or videotaped presentation.
- Can be retaped and edited until the message is just right.
- Can be repeated as often as necessary without burdening the speaker.
- Can be played at the convenience of the audience.
- Can be easily duplicated and given to employees who can then play the tape at home or while driving.
- Can be used as a substitute for an executive who might not handle the presentation well.

The disadvantages of the audiotaped presentation are that it is:

- Not as personal as a verbal presentation.
- Not as personal as a videotaped presentation.
- Less likely than either a verbal or videotaped presentation to stir a response.
- Moreover, audience participation is impossible. Consequently, the benefits of questions, comments, concerns, and employee input are lost.
- Although less is required than with videotapes, making and copying audiotapes still requires production equipment.

===== TSM CASE STUDY =====

What Would You Do?

Andrea Cook isn't really sure what she should do. As public relations director for Microcom, Inc., she knows how to communicate with the public. But communicating a health and safety plan to employees is a different kind of assignment than she is accustomed to. She is supposed to have a *nearly final* communication plan on her boss's desk by noon tomorrow. If Cook asked you to tell her what the plan should consist of and why, what would you tell her?

Newsletter

Organizations that produce a newsletter can make good use of this medium for spreading the word about the safety and health plan. Advantages of this method are as follows:

- Wide distribution to all employees.
- Information about the safety plan can be boxed, carried on the front page, or presented in some other way that calls attention to it.

Newsletters have drawbacks, though. The most important disadvantages of this method are as follows:

- Space limitations do not allow for an in-depth treatment of the subject.
- May not be read by all employees.
- If not presented well, information about the safety and health plan may get lost among other articles.
- Printing deadlines can cause timeliness problems (e.g., the need to wait for the next regularly scheduled edition before a story can be printed).

Special Bulletin or Memorandum

A special bulletin or a memorandum from the CEO is an excellent way to focus the attention of employees on a given subject. This method has the following advantages:

- *Timeliness*. Unlike newsletters, special bulletins and memorandums have no deadlines. Consequently, they can be developed and distributed at any time.
- *Covers a single subject*. There is no chance that information about the safety and health plan will get lost among other stories.
- *No space limitation*. As much information as is necessary can be covered in a special bulletin or memorandum.

In spite of their positive points, special bulletins and memorandums do have their drawbacks. The most important disadvantages of this method are as follows:

- The logistics and cost of production and distribution.
- Many employees simply do not read written material (they suffer from the television syndrome).

Posters and Bulletin Board Notices

One way to put the message before employees everyday is to create posters or notices containing the vision, guiding principles, and broad objectives and display them in heavily trafficked locations and/or on bulletin boards. Advantages of this method are as follows:

- Keeps the message in front of employees on a daily basis.
- Keeps the message in front of customers and suppliers who visit the facility.

- Employees can be involved in the development, display, and periodic changing of the posters/notices.

Despite their positive points, posters/notices also have their drawbacks. The most important of these are as follows:

- If not attractively designed, they will appear as wall clutter and be ignored.
- If not changed out on a regular basis they will begin to blend into the background and be ignored.
- They can be viewed by employees as empty sloganeering if used as the only communication tool.

An excellent way to make sure that every employee knows the vision, guiding principles, and broad objectives is to develop and distribute wallet cards. Their advantages are as follows:

- Employees typically like the cards and, when asked about the organization's plan, will pull the card out in response.
- Handy and readily accessible.
- Good public relations tool (for customers and suppliers).

However, though effective, wallet cards have drawbacks. The most important disadvantages are as follows:

- May require frequent replacement as they wear out or are lost.
- May be put in a wallet and forgotten about.

A creative team of managers and employees could identify many additional methods for informing employees about the safety and health plan. Regardless of the approach chosen by a given organization, using a variety of complementary methods is important.

TSM TIP

Health, Safety, and Competitiveness

"Health and safety contribute to competitiveness in the following ways: (a) by helping companies attract and keep the best people: (b) by allowing employees to focus on peak performance without being distracted by concerns for their health and safety; (c) by freeing money that can be reinvested in technology updates; and (d) by protecting the corporate image."[4]

THE NEED FOR FEEDBACK

Developing the safety and health plan is the job of the TSM Steering Committee. How-ever, this does not mean that there is no place for employee input. Feedback from employees is one of the most important aspects of the communication step. Executive managers—the TSM Steering Committee members—need to know what employees think about the safety and health plan. Is it comprehensive enough? Has anything been overlooked?

Throughout the TSM implementation process, and for as long as TSM is the approach used for ensuring a safe and healthy workplace, the TSM Steering Committee and the IPTs will apply the A-P-D-C-A Cycle (Figure 7–6). The five components of this cycle are *Assess-Plan-Do-Check-Adjust*. No matter what the undertaking in question happens to be, the Steering Committee or the IPT will proceed as follows:

1. Assess the situation and collect all of the relevant facts.
2. Plan in such a way as to correct weaknesses while simultaneously exploiting strengths.
3. Do what is planned (implement the plan).

Figure 7–6
Applying the A-P-D-C-A Cycle

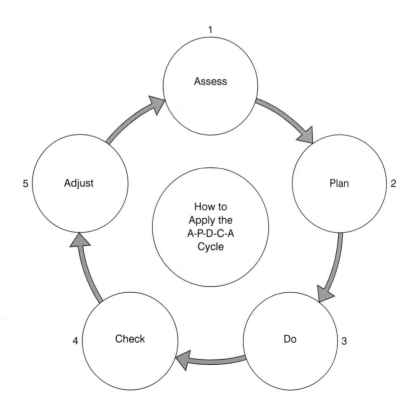

4. Check (monitor and evaluate) to determine if the desired results are being achieved.

5. Adjust as required, including starting over again at the planning stage if necessary.

The TSM Steering Committee applies the A-P-D-C-A Cycle when developing and communicating the safety and health plan. By Step 7, the Steering Committee has assessed the need for a plan, developed one, and is ready to implement it. The *communicate and inform* step is a preliminary *check* before moving into the full implementation of the plan. Informing employees about the plan gives them an opportunity to provide feedback. Questions, comments, opinions, and ideas offered by employees should be recorded by the Steering Committee member who hears them. In addition, Steering Committee members should actively solicit comments, questions, opinions, and ideas from employees. This feedback, too, should be recorded.

Once employee feedback concerning the plan has been collected and recorded, it should be examined by the TSM Steering Committee and the safety and health plan should be adjusted if necessary. Once the A-P-D-C-A Cycle has been completed for the plan, the Steering Committee is ready to move on to the next major phase of the implementation process—*identification and assessment*.

STEP 7 IN ACTION

In anticipation of the TSM Steering Committee completing MPC's safety and health plan, Mack Parmentier had met with the company's public relations director. He knew the committee would expect him to propose a communication plan for informing all of MPC's employees about the safety and health plan; this was a task for which Parmentier had no training and only limited experience. MPC's Director of Public Relations, Jane Morgan, had been exceptionally helpful. She had taken an immediate interest in the project.

In less than two hours, Morgan and Parmentier had worked out a plan. By the end of the day, they had calculated costs and drafted a budget. Now, the more Parmentier reviews the proposed communication plan, the more he likes it and the more he appreciates Jane Morgan's help. Tomorrow, at the TSM Steering Committee's regularly scheduled meeting, Parmentier will present a plan that has five components:

- *Component 1: Special Bulletin.* MPC's CEO will draft a special bulletin containing the complete safety and health plan. In addition, the bulletin will contain an explanation of why the plan was developed and what it means to employees. The bulletin will be distributed to all employees along with an invitation to attend one or two employee meetings in which MPC's CEO will make a verbal presentation about the safety and health plan.

- *Component 2: Large Employee Meetings.* The CEO will make two verbal presentations to which all employees are invited. For convenience, the first presentation will be scheduled at the beginning of the work day; the second at the end of the work day.

- *Component 3: Department-Level Presentations.* Following the CEO's two large employee meetings, each respective member of the TSM Steering Committee will conduct department-level question and answer sessions. During these sessions, each TSM Steering Committee member will record the feedback and concerns of employees for sharing later with the committee.

■ *Component 4: Wall Posters.* Wall posters containing the vision, guiding principles, and broad goals will be developed and displayed strategically throughout the company's facilities. An ongoing poster development program will be conducted by the public relations department in which employees submit their ideas for posters. Winners will receive a cash incentive and posters will be changed about every 60 days.

■ *Component 5: Wallet Cards.* Laminated wallet cards containing the vision, guiding principles, and broad goals will be developed and distributed during the CEO's verbal presentation.

SUMMARY

1. It is important to inform employees about the safety and health plan. A lack of communication can lead to rumors which, in turn, can lead to problems such as unnecessary stress, low morale, panic, anger and hostility, and turnover.

2. Communicating with employees about the safety and health plan while it is being developed can lead to confusion. It is better to wait until the plan has been completed before sharing it with employees. However, it is a good idea to let employees know that a plan is being developed, why, and how it will affect them.

3. When informing employees about the completed safety and health plan, executives should explain the organization's vision, its guiding principles, mission, and broad objectives, as well as why the plan was developed and what the plan means to employees.

4. When communicating with employees about the safety and health plan, it is important to use a variety of methods such as the following: verbal presentations, videotaped presentations, audiotaped presentations, newsletter, special bulletins or memorandums, posters, bulletin board notices, and wallet cards.

5. When implementing TSM, executives and members of IPTs should apply the A-P-D-C-A Cycle continually. This cycle proceeds as follows: (a) assess, (b) plan, (c) do, (d) check, and (e) adjust.

KEY TERMS AND CONCEPTS

Adjust	Check
Anger and hostility	Do
Assess	Low morale
Audiotaped presentation	Newsletter
Bulletin board notices	Panic

Plan   Unnecessary stress
Posters   Verbal presentation
Special bulletin   Videotaped presentation
Turnover   Wallet cards

REVIEW QUESTIONS

1. Explain the rationale for informing employees about the completed safety and health plan.
2. Why is it important to wait until the safety and health plan has been completed before sharing it with employees?
3. When communicating with employees about the safety and health plan, *what* should they be told?
4. List the various methods that might be used to communicate with employees about the safety and health plan. Give one advantage and one disadvantage of each method.
5. Explain the A-P-D-C-A Cycle and how it is used during a TSM presentation.

ENDNOTES

1. David L. Goetsch, *Occupational Safety and Health in the Age of High Technology*, 2nd ed. (Upper Saddle River, N.J.: Prentice Hall, 1996), p. 339.
2. David K. Lindo, "The 10 Commitments of Quality Supervision," *Quality Digest*, Vol. 16, No. 6 (June 1996), p. 41.
3. Goetsch, p. 349.
4. Goetsch, p. 349.

Identify the Organization's Safety and Health Strengths and Weaknesses

===== MAJOR TOPICS =====

- Rationale for Identifying Strengths and Weaknesses
- Who Makes the Determination?
- Capitalizing on Strengths
- Addressing Weaknesses
- Step 8 in Action

In Step 7, the TSM Steering Committee communicated with employees using a variety of methods to inform them about all aspects of the organization's safety and health plan. Step 7 completed the *planning and preparation* phase of the TSM implementation process. Step 8 begins the next phase, *identification* and *assessment*. It involves identifying the organization's strengths and weaknesses from the perspective of TSM.

RATIONALE FOR IDENTIFYING STRENGTHS AND WEAKNESSES

The structure for identifying the organization's strengths and weaknesses from the perspective of TSM can be found in the safety and health plan. More specifically, the necessary structure is imposed by the broad objectives established as part of the plan.

The rationale for identifying strengths and weaknesses grows out of the objectives. If the organization is going to accomplish its objectives, executive managers must begin with a thorough understanding of the organization's strengths and weaknesses relative to the objectives. Otherwise they won't know how to most efficiently and effectively apply limited resources to achieve the objectives.

An organization's strengths can give it competitive advantages if properly exploited. In order to do so, the organization must first identify its strengths. Correspondingly, an organization's weaknesses can become competitive disadvantages unless properly dealt with. The first step in dealing with organizational weaknesses is identifying them.

Figure 8–1 is a list of organizational strengths and weaknesses The Douglas Corporation identified relating to workforce safety and health. A quick examination of the weaknesses identified will show that they tie directly to The Douglas Corporation's broad objectives for safety and health, Figure 8–2.

WHO MAKES THE DETERMINATION?

Ultimate responsibility for determining an organization's strengths and weaknesses relative to safety and health rests with the TSM Steering Committee. However, the committee should solicit input from the broadest possible base of employees at all levels. Employees who perform the work associated with producing the organization's products or delivering its services have detailed knowledge of the hazards they confront daily. That knowledge should be tapped by the Steering Committee. There are several different methods that can be used for soliciting employee input concerning strengths and weaknesses, as illustrated by Figure 8–3. All of these methods revolve around the organization's broad objectives.

Employee Survey

This method involves converting the list of broad objectives into a survey instrument such as the one in Figure 8–4. The survey is distributed to all employees with instructions to return completed instruments to a central collection point. Surveys are com-

Figure 8–1
One Company's Assessment of Its Strengths and Weaknesses

**The Douglas Corporation
Strengths and Weaknesses
Workplace Safety and Health**

Strengths

- Executive commitment by executive management
- An active and strong TSM Steering Committee (executive management team)
- A well educated/experienced safety director
- A comprehensive safety and health plan

Weaknesses

- Numerous old machines without proper machine guards
- Old facilities in which the electrical wiring may be overloaded
- Insufficient storage and disposal capabilities for toxic substances
- Several new automated machines for which safety procedures have not yet been developed

**The Douglas Corporation
TSM Steering Committee
Safety and Health Broad Objectives**

- Establish a work environment that is free of ergonomic hazards.
- Establish a work environment that is free of fall/acceleration-related hazards.
- Establish a work environment that is free of lifting hazards.
- Establish a work environment that is free of heat/temperature hazards.
- Establish a work environment that is free of pressure hazards.
- Establish a work environment that is free of electrical hazards.
- Establish a work environment that is free of fire hazards.
- Establish a work environment that is free of toxic substance hazards.
- Establish a work environment that is free of explosives-related hazards.
- Establish a work environment that is free of radiation hazards.
- Establish a work environment that is free of noise hazards.
- Establish a work environment that is free of vibration hazards.
- Establish a work environment that is free of automation-related hazards.
- Establish a work environment that is free of bloodborne hazards.

Figure 8–2
That Representative Company's Objectives

pleted anonymously and summarized by the TSM Facilitator. The TSM Steering Committee receives only the summary document. The anonymous approach will increase the amount and improve the quality of the input received from employees.

Small Group Meetings

Small group meetings are conducted by the TSM Facilitator. This individual distributes the same instrument used in the employee survey (Figure 8–4) to all participants in the meetings. The instrument is used to give structure to the discussion that takes place.

TSM TIP

Workers' Compensation and Stress

An often-stated objective of workers' compensation legislation is the reduction of expensive and time-consuming litigation. However, many cases—particularly those involving stress claims—still go to court.

Figure 8–3
Means of Soliciting Employee
Input

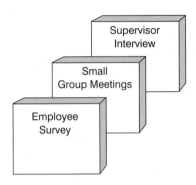

The TSM Facilitator, taking one broad objective at a time, begins a dialogue concerning weaknesses. Participants are encouraged to verbalize their input, and discuss their opinions. This type of interaction will accomplish two goals. First, discussion will prod the memories of participants concerning weaknesses they might have forgotten, or to which they have become accustomed. Second, it will tend to draw out employees who are reluctant to speak up.

The TSM Facilitator acts as a recorder, making note of any and all weaknesses brought up that receive group consensus. Once the weaknesses have been identified, the process is repeated for strengths with the TSM Facilitator again acting as recorder. The final list of weaknesses and strengths is presented to the TSM Steering Committee. This method is an excellent way to solicit comprehensive, detailed input from employees. However, it can be a time-consuming approach in large organizations.

Supervisor Interviews

With this method, individual supervisors interview all employees who report directly to them. Supervisors use the employee survey as a guide. Input is recorded and forwarded to the TSM Facilitator who analyzes and summarizes the input collected by each supervisor. The summary is presented to the TSM Steering Committee. This method has the advantage of delegation. Consequently, it is popular in large organizations. However, it can limit the quantity and quality of input if employees are reluctant to be open with their supervisors.

Once employee input concerning strengths and weaknesses has been collected, the TSM Steering Committee reviews what has been learned. Steering Committee members should understand that the list of weaknesses collected from employees is typically longer than the list of strengths. This is to be expected and should not be viewed in a negative light.

There are two reasons why the list of weaknesses is typically longer than the list of strengths. First, employees—being human—will tend to focus more on problems in their work environment than on positive conditions. Positives are often taken for granted by employees. After all, a hazard-free work environment is not a gift from man-

Safety and Health
Strengths and Weaknesses Survey

Directions

The TSM Steering Committee is in the process of identifying our company's strengths and weaknesses in the area of workplace safety and health. As an employee who is close to our work processes, you can provide invaluable information to the committee by completing this survey instrument. Each entry in the survey is one of our company's broad objectives for safety and health. Please note any weaknesses or strengths in the space provided after each objective.

- To have a workplace that is free of ergonomic hazards.

Strengths	Weaknesses
______________________	______________________
______________________	______________________
______________________	______________________
______________________	______________________
______________________	______________________

- To have a workplace that is free of fall/acceleration-related hazards.

Strengths	Weaknesses
______________________	______________________
______________________	______________________
______________________	______________________
______________________	______________________
______________________	______________________

Figure 8–4
Sample Survey Instrument

agement. It's how the workplace is *supposed* to be. Second, there may actually be more negatives than positives. If this were not the case, would executives be so interested in implementing TSM? Probably not.

═══════════ TSM CASE STUDY ═══════════

What Would You Do?

Although John Pace has been a safety professional for more than 10 years, TSM is a new concept to him. As the TSM Facilitator for Baker Products, Inc., Pace is having to learn on the run, but so far he has been able to stay one step ahead in the implementation process.

The next challenge Pace must contend with is identifying his organization's strengths and weaknesses from a safety and health point of view. Baker Products, Inc, is a small company with just 136 employees, all located in one facility. If you were John Pace, how would you go about identifying this organization's strengths and weaknesses?

CAPITALIZING ON STRENGTHS

A baseball team that has an outstanding power hitter tries to give him as many at-bats as possible during every game. A basketball team that has a high-percentage three-point shooter feeds him the ball at every opportunity. A football team that has a breakaway running back will give him the ball 20 or more times every game. These teams are capitalizing on their strengths; something every winning team does and does well. It's one of the reasons they win consistently. This same concept—capitalizing on strengths—also applies to organizations trying to compete in the global marketplace.

To understand how this concept can be applied in improving the work environment, examine the strengths set forth in Figure 8–1. How might an organization go about making the most of strengths such as these? The first strength noted is *executive commitment* to TSM. Executive commitment is one of the keys to an organization's success in any endeavor. To capitalize on executive commitment, the organization should strike while the iron is hot. In other words, it should complete the TSM implementation process with dispatch while the level of executive commitment is at its peak.

The second strength noted is *an active and strong TSM Steering Committee*. The best way to capitalize on this strength is to translate commitment into resources and proactive decisions. The TSM Steering Committee, if it consists of the organization's executive managers, has the authority to commit the resources necessary to ensure a safe and healthy workplace. It also has the authority to make the decisions necessary to get TSM fully implemented over a relatively short period of time.

The third strength noted in Figure 8–1 is *a well-educated, experienced safety director*. The organization's safety director is a professional in the field. He or she has studied safety and health in college, in technical school, or through corporate training opportunities, and has gained practical experience on the job. The best way to capitalize on this

---

TSM TIP

Causes of Workplace Stress

Some of the more common causes of workplace stress are as follows: reorganizations, buyouts, layoffs, mandatory overtime, workload variations, work pace, insufficient opportunities for advancement, bureaucracy, low pay, outdated technology, insufficient staff, and shift rotations.[1]

---

strength is to name this individual the organization's TSM Facilitator, and give him or her the latitude and support needed to fully implement the concept.

The final strength noted in Figure 8–1 is *a comprehensive safety and health plan.* This means the organization's executive managers already have identified a vision, mission, guiding principles, and broad objectives relating to safety and health. The best way to capitalize on this strength is to strategically implement the safety and health plan.

ADDRESSING WEAKNESSES

Every organization has weaknesses, even the best. However, one of the key differences between organizations that win consistently and those that don't is how they address their weaknesses. Some organizations ignore their weaknesses, hoping to overcome them by capitalizing on their strengths. These organizations are like the baseball team that has several strong hitters, but a weak pitching staff. Rather than strengthen its pitching staff, the team expects its strong hitters to produce enough runs to compensate for the weak pitching. Of course, such a team will have an inconsistent record at best. But consider what might happen if the team did what was necessary to strengthen its pitching staff.

What if the team traded for a couple of strong pitchers, taught some new pitches to its existing staff, changed the pitching rotation, and converted an aging starter or two into relief pitchers? In just a short while the team might find that with better pitching, it

---

TSM TIP

Machine Safeguards

Machine safeguards minimize the possibility of machine operator accidents. Such accidents are often the result of an individual making contact with a machine component, flying metal chips, chemical or hot metal splashes, stock kickbacks, or mechanical malfunction.

can win with fewer runs. Now, when the team produces five or more runs—something it can do consistently—it finds itself winning instead of losing to teams that in the past were able to produce even more runs off of its weak pitchers.

The team in this example converted itself from a loser to a consistent winner by addressing its weaknesses while simultaneously capitalizing on its strengths. Organizations that want to create a worldclass work environment in which employees can achieve consistent peak performance should follow this same course: capitalize on strengths while simultaneously addressing weaknesses.

In a TSM setting, weaknesses are addressed by IPTs. Weaknesses addressed in this step will be assigned to IPTs in Step 12. For now they are kept on file by the TSM Steering Committee. Having information about weaknesses early in the overall implementation process gives the TSM Steering Committee time to consider the financial ramifications of correcting the weaknesses. It also gives the TSM Facilitator time to consider the order in which weaknesses should be assigned to IPTs, and to recommend these priorities to the TSM Steering Committee.

STEP 8 IN ACTION

Even though he had prepared the executives beforehand, Mack Parmentier could tell that the list of weaknesses identified by MPC's employees had shocked the members of the TSM Steering Committee. Although he hadn't said so, Parmentier had been surprised himself. MPC's work environment was apparently in even worse shape than he had thought.

Based on his recommendation, the TSM Steering Committee had decided to use an employee survey followed by small group meetings conducted by Parmentier. The survey had been effective at identifying numerous weaknesses, but in several instances, employee input either wasn't clear or didn't provide sufficient detail. The small group meetings had allowed Parmentier to question, probe, and seek clarification concerning input that wasn't entirely clear from his analysis of survey forms.

Combining an employee survey with small group meetings had worked so well that Parmentier was now convinced that this was the way to go. He was scheduled to make a presentation on TSM at an upcoming meeting of occupational safety and health professionals. He decided to mention the combination approach during his presentation. Even

---

TSM TIP

Head Injuries on the Job

Head injuries on the job have been the driving force behind the development of tougher hard hats. However, even with today's durable headgear, more than 120,000 people sustain head injuries on the job every year.

with his experience, Parmentier had been surprised at how forthcoming employees had been in a small group setting away from their supervisors and managers.

In addition to providing more depth and detail concerning weaknesses that should be addressed, the small group meetings had given Parmentier an opportunity to ask employees to critique the survey form he had developed. Based on their input, he planned to make several changes to the form. But one thing that would not change was the anonymity aspect. There had been strong concerns among employees that without anonymity they would be less forthcoming with their input, especially when it came to identifying supervisors who push employees to neglect safety procedures.

SUMMARY

1. If an organization is going to accomplish the broad objectives in its safety and health plan, executive managers must begin with a thorough understanding of the organization's strengths and weaknesses relative to the objectives.

2. Ultimate responsibility for determining an organization's strengths and weaknesses relative to safety and health rests with the TSM Steering Committee. However, the committee should solicit input from a broad base of employees.

3. Several different methods can be used to solicit employee feedback about an organization's safety strengths and weaknesses, each having its pros and cons. They include the employee survey, small group meetings, and supervisor interviews.

4. Organizations capitalize on strengths by taking advantage of them and exploiting them for maximum benefit. They address weaknesses by identifying them and doing what is necessary to make the improvements that are called for.

KEY TERMS AND CONCEPTS

Addressing weaknesses   Small group meetings

Capitalizing on strengths   Supervisor interviews

Employee survey

REVIEW QUESTIONS

1. Why is it important for an organization to identify its strengths and weaknesses relative to safety and health.?

2. Explain how the determination of an organization's strengths and weaknesses relative to safety and health should be made.

3. Describe each of the following methods for identifying strengths and weaknesses. Include the relative advantages and disadvantages of each.
   - Employee survey

- Small group meetings
- Supervisor interview

4. How does an organization go about capitalizing on its strengths?

5. How might an organization address its weaknesses?

ENDNOTES

1. L. Cope, "Quiz Developed to Determine Workplace Stress," *Tallahassee Democrat*, June 2, 1991, pp. 1E, 3E.

Identify Safety and Health Advocates and Resisters

- Rationale for Identifying Advocates and Resisters
- Characteristics of Advocates and Resisters
- Cautions to Observe
- Advocates and Resisters in the Initial Stages of TSM
- Step 9 in Action

With Step 8 completed, the organization has a thorough understanding of its strengths and weaknesses relative to safety and health. Eventually, much of the organization's attention will be focused on improving the weaknesses identified in Step 8 and others that are identified subsequently—a process that will continue forever.

Although it would seem that all employees at all levels would be advocates of such a process, this is not always the case. In every organization there are both advocates and resisters of innovation. Advocates will do what they can to promote a successful implementation. Resisters will seek to block, either actively or passively, the innovation in question. Those who feel threatened by the innovation will actively—although not necessarily openly—oppose it. People who resist because they do not like new ideas or change will promote failure through passive means such as a lack of effort. This step involves identifying the employees who are likely to support TSM (its advocates) and those who are likely to oppose it (the resisters).

RATIONALE FOR IDENTIFYING ADVOCATES AND RESISTERS

In the early stages of implementation, an organization's commitment to TSM—or any other innovation—is typically fragile at best. Consequently, it is especially important that early-stage TSM activities run smoothly. Problems and failures experienced early in the process will be magnified, even in the eyes of supporters. Resisters will use early-stage problems to fan the flames of doubt, particularly with marginal supporters who are especially susceptible to such tactics. A false start in the implementation of a new concept such as TSM can be difficult or impossible to overcome. For this reason, it is important to proceed carefully at the beginning of the implementation, doing everything possible to limit problems and promote success. An effective strategy in this regard is to identify advocates and resisters of TSM, and, at least in the early stages, enlist only advocates in the various implementation activities undertaken.

Identifying advocates and resisters is a Steering Committee function. The first step in the process is to list all management and supervisory personnel in the organization. Then, Steering Committee members brainstorm concerning the likely reaction of each individual on the list. Typically, the members of the Steering Committee will know these individuals well enough to draw a reasonable conclusion as to where each will initially stand on TSM. The TSM Facilitator can be helpful at this point because he or she will probably have some insight into the attitudes of supervisors and managers toward safety and health in general. As the organization's principal safety and health professional, the TSM Facilitator will usually know which managers and supervisors encourage adherence to safety procedures, and which turn their backs when procedures are ignored.

The process is hardly scientific. Rather, it involves making an educated guess as to how people will react to change. When no member of the Steering Committee has a feel for how a given individual on the list will react, the individual's general reaction to change can be used as an indicator. A good rule of thumb is to err on the safe side. If a person cannot be categorized as an advocate with a comfortable degree of certainty, that person should be categorized—at least for the time being—as a resister.

Identifications made at this point are temporary only. Later in the process it will become evident who the true advocates and resisters are. Of course, as the implementation proceeds, individuals who are originally identified as resisters will have ample opportunities to become advocates. As TSM begins to produce results, managers and supervisors who originally resisted the concept may become some of its most ardent supporters. This happens frequently, not just with TSM, but with workplace innovations in general. In organizations that have experienced one or two years of success with an innovation, it is usually difficult to find any remaining detractors.

Once the list of managers and supervisors has been divided into resisters and advocates, the process is repeated for employees. Steering Committee members may know less about line employees than they know about managers and supervisors. Conse-

---

TSM TIP

Heat Burn Injuries

Approximately 40 percent of all heat burn injuries in the workplace occur in a manufacturing setting. The most common causes are flame, molten metal, petroleum asphalts, steam, and water.[1]

quently, the identification process can be more difficult. Again, the TSM Facilitator—as the organization's chief safety and health professional—may be able to provide some insight. The next section explains strategies the Steering Committee can use for categorizing line employees as advocates and resisters.

CHARACTERISTICS OF ADVOCATES AND RESISTERS

Advocates and resisters, regardless of the innovation in question, have a number of observable characteristics. These characteristics can provide valuable insight concerning how a given employee will react to innovation, new ideas, and change. Figure 9–1 shows the types of characteristics that are typical of employees who are usually positive change agents (advocates) in the workplace. These characteristics are explained as follows:

- *Innovator.* Some employees like to try new methods and new approaches for doing their jobs better. Such employees are constantly trying new approaches, making modifications, and seeking new procedures. These employees are innovators. Innovative employees are likely to be advocates of TSM.

- *Can-do attitude.* Some employees have a *can-do* attitude about their work. No matter how rushed the schedule or how difficult the challenge, they typically roll up their sleeves, and pitch in with a positive attitude that says, *"We can do this."* Employees with a can-do attitude are likely to be advocates of TSM.

Figure 9–1
Characteristics of Advocates

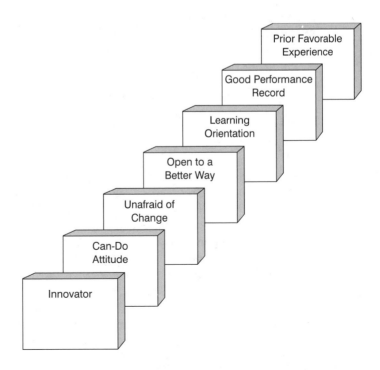

■ *Unafraid of change.* Some employees face change head on. They don't fight it or try to avoid it. They simply get a fix on where the new direction is taking the organization and try to head in that direction themselves. Such employees don't allow themselves the luxury of becoming overly comfortable with existing personnel, policies, procedures, or technologies because they know that all of these things will change continually. Employees who are unafraid of change might be advocates of TSM.

■ *Open to a better way.* Some employees are willing to listen when a manager, supervisor, or fellow employee suggests a better way of doing the job at hand. Rather than being offended or fighting the suggestion, they listen, observe, and objectively draw their conclusions. If the suggestion appears to be valid, they try it. If it proves to be valid, they adopt it. Employees who are open to suggestions about a better way to do the job are likely to be advocates of TSM.

■ *Learning orientation.* Some employees are constantly trying to learn. They want to learn anything and everything that will improve their performance on the job and their potential for advancement. They read professional and technical literature, attend seminars, take night courses, and observe people who are known to be *masters* in a given field. Employees with a learning orientation are likely to be advocates of TSM.

■ *Good performance record.* Some employees constantly perform at peak levels. Day after day they can be counted on to meet or exceed performance expectations and to help others follow suit. Employees with a good performance record are likely to be advocates of TSM, particularly if they see it as a way to improve performance.

■ *Prior favorable experience.* Employees who have worked in an organization for more than a year have experience in responding to change. Consequently, they have a record—if only in the minds of management personnel—of responding. Some employees respond in a positive way to change and become agents of it. Such employees are likely to be advocates of TSM.

It is probably overly optimistic to state with certainty that an employee who has just one of these characteristics will be an immediate advocate of TSM. However, an employee who has two or more of these characteristics can—with a comfortable level of assurance—be categorized as an advocate. If the initial classification is wrong, the employee's true attitude will show through soon enough.

Figure 9–2 shows the types of characteristics commonly displayed by employees who typically resist innovation or change. These characteristics are as follows:

■ *Can't do attitude.* Some employees have a *can't do* attitude. No matter what kind of new idea is proposed, they will find reasons why it cannot be done. Can't-do employees are quick to turn a minor challenge into an insurmountable obstacle. Their grab bag of excuses can be inexhaustible, but what quickly becomes apparent is that every respective excuse is just another way of saying, "I don't like change." Employees with a can't do attitude will probably resist TSM.

■ *Status-quo attitude.* Some employees have a tendency to get comfortable with the way things are in their lives, even when the way things are isn't particularly good.

Figure 9–2
Characteristics of Resisters

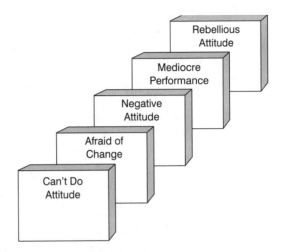

Such employees have a status-quo attitude. They like to keep things the way they are now, the way they are accustomed to. Status-quo employees would rather accommodate a less than desirable situation than deal with the change required to improve it. For such people, familiarity is a high priority. Status-quo employees will probably resist TSM.

■ *Afraid of change.* Some employees can become so wedded to the status quo that they actually fear change. Such people form a strong attachment to what is known and what is familiar in their lives. They are the type of people who subscribe to such folk wisdom as "The devil you know is better than the devil you don't know." Employees who are afraid of change will probably resist TSM.

■ *Negative attitude.* Some employees have a negative attitude toward their work or toward work in general. These types of employees typically fall into one of two categories. The first category consists of employees who dislike their jobs, or just dislike working in general. Because of this, they develop a negative attitude toward anything having to do with their jobs specifically and with work in general. The second category consists of people who have negative attitudes about life in general. An employee with a negative attitude will probably resist TSM. However, a caveat is in order here. If the cause of the negative attitude is an unsafe/unhealthy workplace, the employee may turn out to be a staunch advocate of TSM.

■ *Mediocre performance.* All organizations, even those with worldclass performance records, have employees who do a mediocre job. Mediocre performance that is the result of insufficient training, coaching, and mentoring is the fault of the organization. Mediocre performance that occurs in spite of training, coaching, and mentoring is the employee's fault. An employee who, even with the proper developmental assistance and management support, still performs at a mediocre level will probably resist TSM. Those whose mediocre performance is attributed to a bad attitude, insufficient ambition, or a lack of interest will resist because, for them, TSM repre-

---

TSM TIP

Human Factors Theory of Accident Causation

"The Human Factors Theory *of accident causation attributes accidents to a chain of events ultimately caused by human error. It consists of the following three broad factors that lead to human error: overload, inappropriate response, and inappropriate activities."*[2]

---

sents one more challenge they don't want to deal with. Those who perform at a mediocre level because this is the best they can do will resist TSM because they are fighting every day just to keep their heads above water. For such employees, anything new might elevate the job to a level that is completely out of reach.

■ *Rebellious attitude.* Some employees take their negative attitudes to the level of rebelliousness. Such employees are typically anti-management in their thinking regardless of the issue in question. Why such employees are allowed to remain on the payroll is a good topic for another discussion. For the purposes of TSM, rebellious employees should be categorized as resisters, at least in the early stages.

Using Known Characteristics to Identify Advocates and Resisters

It may be impossible in all but the smallest companies for executive managers to know line employees well enough to predict with a comfortable degree of accuracy how they will react to TSM. However, this does not mean that line employees cannot be categorized as potential advocates and resisters. By matching the characteristics of resisters from Figure 9–2 with the known characteristics of line employees, reasonable conclusions can be drawn concerning probable employee attitudes.

The Steering Committee, with the help of the TSM Facilitator, has essentially three methods for doing this job at its disposal, as shown in Figure 9–3. Each of them will be discussed next.

Steering Committee Brainstorming

The Steering Committee, working with a list of all line employees, can categorize any who are known. A brainstorming session in which employees are discussed from the perspective of the characteristics of advocates and resisters can be helpful. This approach has the advantage of confining the advocates/resisters issue to the Steering Committee. This is the best way to ensure that employees who are initially categorized as resisters are not permanently stigmatized. This issue is dealt with later under the heading Cautions to Observe. The brainstorming approach has the disadvantage of limited effectiveness. In all but the smallest organizations, executives will typically have only a limited knowledge of line employees to work with.

Figure 9–3
Methods for Identifying
Advocates and Resisters

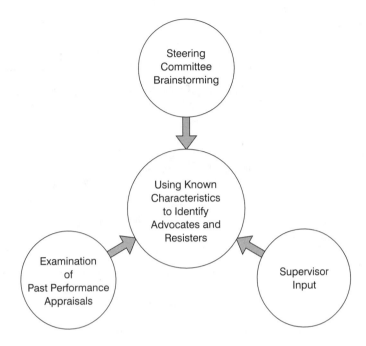

Supervisor Input

The Steering Committee can ask the organization's supervisors to categorize those who report directly to them, based on the known characteristics of advocates and resistors. This method has two advantages. First, it is probably the fastest way to complete the process of categorizing line employees as potential advocates and resisters. Second, supervisors will be more knowledgeable than executives concerning the attitudes of line employees.

Although these advantages are attractive, this method has several disadvantages that lessen its viability. First, the supervisors in question may themselves be resisters when it comes to implementing TSM. It makes no sense to ask supervisors who are resisters to make judgements about employees concerning their potential reactions. Second, the more people who are involved in categorizing employees the more likely it becomes that employees will be permanently stigmatized if they are categorized as resisters. Finally, this method is sure to find its way into the organization's grapevine in the form of rumors. Rumors, it is known, follow a predictable and often harmful course: Phase 1—Initiation, Phase 2—Magnification, and Phase 3—Multiplication.

Examination of Past Performance Appraisals

The types of characteristics illustrated in Figures 9–1 and 9–2 often show up in performance appraisals in one form or another. Well-designed appraisal forms contain criteria that measure employee performance behavior relative to the types of characteristics in question. Examples of these kinds of criteria are shown in Figure 9–4.

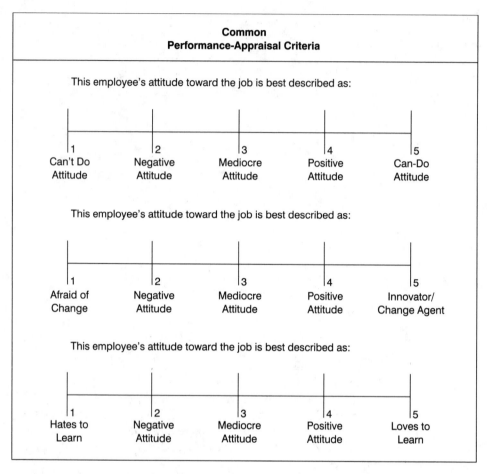

Figure 9–4
Examples of Performance-Appraisal Criteria That Can Be Used in Identifying Advocates and Resisters

Even when such criteria are not part of the appraisal form, indicators relating to the characteristics in question can often be found in the Comments section of the form. The method being described now, that of searching performance appraisal forms for relevant clues, has the advantage of confining the categorization process to a small circle of participants (e.g., the Steering Committee, the TSM Facilitator, and a representative of the organization's human resources department). Limiting the circle of participants, in turn, limits the need for rumor control as well as limiting the likelihood that a temporary classification as a resister will become a permanent stigma. On the other hand, this method has the disadvantage of being time consuming.

TSM TIP

Requirements of Safeguards

Safeguards can be designed to protect employees from harmful contact with machines without inhibiting work processes. The National Safety Council has established the following requirements for safeguards:[3]

- *Prevent contact*
- *Be secure and durable*
- *Protect against falling objects*
- *Create no new hazard*
- *Create no interference*
- *Allow safe maintenance*

CAUTIONS TO OBSERVE

It is critical in the early stages of a TSM implementation to involve employees who will help the concept catch on and succeed. Correspondingly, it is equally critical to avoid the involvement of employees who will drag the concept down. Neither of these objectives can be accomplished without first making some judgements about how individual employees will respond to TSM. Categorizing employees as *initial* advocates or resisters is a necessary activity. It is also a potentially dangerous activity.

If the categorization process is mishandled, it can create problems that will persist for years. Consequently, it is critical that appropriate cautions be observed by all participants in the process. These cautions are as follows (see Figure 9–5):

- Understand that the designation (advocate or resister) initially assigned to employees is temporary. A conscious effort must be made to avoid stigmatizing employees based on their likely reaction to TSM. Participants should be reminded that an employee can be a peak performer and still be a staunch resister of change.

- Keep the number of participants in the categorization process as small as possible. Employees will be concerned, and understandably so, if they learn that they are being classified as advocates or resisters. A rule-of-thumb to remember is that the larger the number of participants in any endeavor, the more difficult it is to keep a lid on it.

- When there is insufficient evidence to make an accurate prediction concerning an employee's probable response to TSM, err on the safe side. It is better to hold potential advocates out of initial activities than to involve resisters in them.

Figure 9–5
Cautions to Observe When
Identifying Advocates and
Resisters

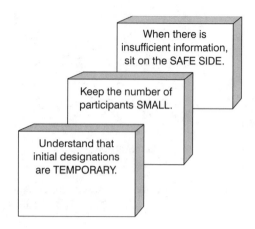

When there is insufficient information, sit on the SAFE SIDE.

Keep the number of participants SMALL.

Understand that initial designations are TEMPORARY.

If these cautions are carefully observed, the TSM implementation will have a better chance to catch on. Once this happens, most initial resisters, including the passive wait-and-see types, will become advocates. This is human nature. Over time, the true attitudes of employees will surface. Consequently, if an advocate is initially misdiagnosed as a resister, time will take care of it. If there are those who remain resisters after TSM has become the norm, they should be dealt with through regular employee processes such as performance appraisals.

No organization can or should tolerate behavior that is contrary to its vision and guiding principles. However, employees who resist a concept after it has become part of the organization's culture typically leave on their own.

TSM CASE STUDY

What Would You Do?

As CEO of ICC Company, Morris Drew is concerned about the categorization process. He is afraid that employees are going to be stigmatized. He is also afraid employees will hear about the process and respond negatively. Mike Bowers, ICC's TSM Facilitator, has been asked to speak to Drew's fears. If you were Bowers, what would you tell Morris Drew?

STEP 9 IN ACTION

Step 9 had presented Mack Parmentier with an interesting challenge. As a new employee of MPC with no prior knowledge to draw on, categorizing his colleagues as advocates or resisters had not been easy. Categorizing managers and supervisors had been less of a problem than categorizing employees. The various TSM Steering Committee members knew MPC's mid-managers and supervisors well enough to draw reasonable conclusions concerning their individual attitudes toward TSM. But line employees had been a problem. Neither the Steering Committee members nor Parmentier knew MPC's line employees well enough to predict how they would react to TSM.

After discussing various employees without being able to draw any reliable conclusions, MPC's CEO had asked Parmentier for a recommendation. Parmentier had asked the Steering Committee to let him work with MPC's Human Resources Office to complete Step 9. His idea was to ask HRM personnel to examine the performance appraisal files of MPC's employees giving special attention to indicators that might be evidence of a can't do attitude; status quo attitude; fear of change; negative attitude toward work, management, and/or fellow employees; mediocre performance; or a rebellious attitude.

Employees whose performance appraisals contained indications of one or more of these characteristics were listed in one column. All other employees were listed in another. The lists were given to the TSM Steering Committee. Employees on the *negative-indicators* list were tentatively coded as potential resisters. To check the veracity of this approach, two supervisors who had already been coded as advocates were asked to code workers in their departments. The two lists were then compared.

Parmentier had been pleased to see that, with one exception, the supervisors' codes matched those on the lists produced by the company's HRM Department. On the HRM list he appeared as a resister. Just to be safe, the Steering Committee decided to leave this employee in the *resister* column.

Once Step 9 was completed, one of the TSM Steering Committee members had asked Parmentier why he hadn't saved time by simply asking all supervisors to code their staff members as either advocates or resisters. The executive had reasoned that if this approach worked well with the two supervisors Parmentier had involved in the process, it should work equally well with all supervisors.

Parmentier had explained that involving all supervisors in the coding process could create more problems than it would solve. Not all supervisors were themselves coded as advocates. Asking supervisors who are resisters to decide whether their people are likely to be advocates or resisters makes no sense. In addition, involving all supervisors just increases the likelihood that employees will be permanently tainted by the resister designation.

Looking back, Parmentier is sure that the Steering Committee is now ready to move on to Step 10—benchmarking initial employee attitudes concerning the work environment.

SUMMARY

1. The rationale for identifying advocates and resisters is to give the TSM implementation a chance to succeed. Involving resisters in early-stage implementation activities is the best way to ensure failure.

2. Characteristics common to advocates of innovation and change are as follows: innovator, can-do attitude, unafraid of change, open to a better way, learning orientation, good performance record, and prior favorable experience.

3. Characteristics common to resisters of innovation and change are as follows: can't-do attitude, status-quo attitude, afraid of change, negative attitude, mediocre performance, and rebellious attitude.

4. The TSM Steering Committee has the following methods at its disposal for categorizing line employees as advocates and resisters: Steering Committee brainstorming, supervisor input, examination of past performance appraisals.

5. Cautions to observe when categorizing employees as initial advocates or resisters are as follows: (a) understand that initial designations are temporary; (b) keep the number of participants small; and (c) err on the safe side.

KEY TERMS AND CONCEPTS

Advocates	Negative attitude
Can-do attitude	Open to a better way
Can't-do attitude	Prior favorable experience
Good performance record	Rebellious attitude
Innovator	Resisters
Learning orientation	Status-quo attitude
Mediocre performance	Unafraid of change

REVIEW QUESTIONS

1. Explain the rationale for identifying advocates and resisters.
2. List and briefly explain the characteristics that are common to advocates of innovation.
3. List and briefly explain the characteristics that are common to resisters of innovation.
4. Explain how to use known characteristics of advocates and resisters for categorizing line employees.
5. What cautions should be observed when initially categorizing advocates and resisters?

ENDNOTES

1. National Safety Council, *Accident Facts* (Chicago: 1993), p. 40.
2. David L. Goetsch, *Occupational Safety and Health in the Age of High Technology*, 2nd ed., (Upper Saddle River, N.J.: Prentice Hall, 1996), p. 36.
3. National Safety Council. *Guards: Safeguarding Concepts Illustrated*, 5th ed. (Chicago: 1987), p. 1.

Benchmark Initial Employee Perceptions Concerning the Work Environment

Employee perceptions concerning the state of the work environment can affect both morale and performance. Since continual improvement of the work environment is a cornerstone of the TSM philosophy, it is important to understand where those perceptions are before improvement begins. Without this knowledge there is no benchmark from which to measure progress. A benchmark, in this sense, is a starting point that is used as the basis for measuring progress in making improvements; for example, a beginning weight for a person trying to lose a few pounds, or a point of departure for a person trying to work up to a 10-mile run.

This step describes the process of benchmarking initial employee attitudes concerning the work environment. Included are the *whys*, *hows*, and *cautions* relating to benchmarking.

RATIONALE FOR ASSESSING ATTITUDES

Picture the following conversation:

"Hi, John. Long time no see. You're looking fit. Have you lost weight?"

"I sure have!" replies John proudly. "I've been on a diet and fitness program for three months. Thanks for noticing."

"How much weight have you lost?"

TSM TIP

Chemical Injuries in the Workplace

The chemicals that are most frequently involved in workplace injuries are acids and alkalines: soaps, detergents, and cleaning compounds; solvents and degreasers; calcium hydroxide; potassium hydroxide; and sulfuric acid.[1]

John, looking slightly confused, hesitates for a moment before answering, "I don't know. I forgot to weigh at the beginning of the program."

John, in this example, knows he is losing weight. He can see it in the mirror, and feel it in the fit of his clothing. But John has no idea how much weight he has lost because he forgot to weigh at the beginning of his diet and fitness program. John has a general impression of weight loss, but he has no hard data concerning how much. If his goal was to lose 20 pounds, how will John know when he has achieved it?

This same concept of needing to have hard data to use in measuring progress also applies to assessing initial employee attitudes. One of the objectives of TSM is to improve employee perceptions/attitudes continually and forever concerning the quality of the work environment. In order to measure progress made in this regard, it is necessary to have a benchmark to work from. The benchmark is the initial starting point. This is the rationale for assessing employee attitudes. Before beginning to make improvements that will, in turn, improve employee attitudes, organizations need to know current employee perceptions. It is important to conduct the initial assessment before the first IPT is put to work identifying ways to improve the quality of the workplace. Then, as workplace improvements are made, employee attitudes/perceptions can be measured periodically. The data from these measurements can be compared to the benchmark that was established initially.

HOW TO ASSESS PERCEPTIONS

There are several ways to assess employee perceptions concerning the quality of the work environment. The most widely used methods are shown in Figure 10–1. Each of these methods has its advantages and disadvantages as described in the following subsections.

Survey (Internal)

The most widely used method for assessing employee perceptions as they relate to the work environment is the employee survey conducted as an in-house project. With this method a survey instrument such as the one shown in Figure 10–2 is developed by the TSM Facilitator. The actual instrument is tailored specifically for the organization in question.

The survey instrument is distributed to all employees with a cover memorandum explaining its purpose, the time frame within which it should be completed, and how con-

Figure 10–1
Assessing Employee Perceptions

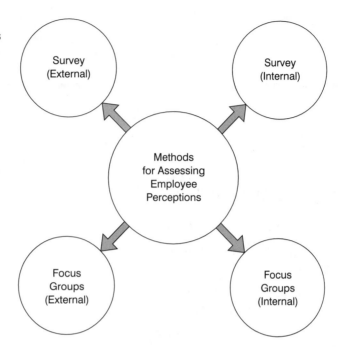

fidentiality will be maintained. A sample of such a memo is shown in Figure 10–3. Once all responses have been collected, they are summarized and the individual forms are destroyed. The internal survey has several advantages. The most important of these are as follows:

■ Involves all employees
■ Can be administered in a relatively short period of time
■ Inexpensive

The internal survey also has disadvantages, the most important of which are as follows:

■ Summarizing responses can be a major undertaking in companies with a large number of employees.
■ Employees may not respond if they don't trust management to maintain strict confidentiality.
■ Does not allow employees to give in-depth responses.

Focus Groups (Internal)

Another widely used method for assessing employee perceptions as they relate to the work environment is the focus group conducted as an in-house project. With this method, employees representing all departments and units within the organization are

Employee Perceptions Survey

Perception Statement	Strongly Disagree	Disagree	Agree	Strongly Agree
1. The workplace is free of unreasonable stress.	_____	_____	_____	_____
2. The workplace is free of mechanical hazards.	_____	_____	_____	_____
3. The workplace is free of falling and impact hazards.	_____	_____	_____	_____
4. The workplace is free of lifting hazards.	_____	_____	_____	_____
5. The workplace is free of heat/temperature hazards.	_____	_____	_____	_____
6. The workplace is free of electrical hazards.	_____	_____	_____	_____
7. The workplace is free of fire hazards.	_____	_____	_____	_____
8. The workplace is free of toxic substance hazards.	_____	_____	_____	_____
9. The workplace is free of explosive hazards.	_____	_____	_____	_____
10. The workplace is free of radiation hazards.	_____	_____	_____	_____
11. The workplace is free of noise/vibration hazards.	_____	_____	_____	_____
12. The workplace is free of automation hazards.	_____	_____	_____	_____
13. The workplace is free of bloodborne pathogen hazards.	_____	_____	_____	_____
14. The workplace is free of ergonomic hazards.	_____	_____	_____	_____
15. The workplace is free of violence hazards.	_____	_____	_____	_____

Figure 10–2
Assessing Employee Perceptions

invited to be members of a focus group. In larger organizations, more than one group may be required because the size of a group should be between 10 and 15. If it is smaller, the group may represent too narrow a perspective. If larger, the group may be unwieldy.

As with the survey method, a document such as the one in Figure 10–2 is developed by the TSM facilitator. This document is used to prompt discussion while simultaneously focusing the talk on concerns about the work environment. Focus group meetings can become gripe sessions unless a strong facilitator keeps the discussion on track. The

---

Freemont Mining & Minerals, Inc.

Memorandum

TO: All Employees
FROM: Jackson Drew, TSM Facilitator
SUBJECT: Employee Perceptions Survey

Attached you will find a survey instrument and an interoffice envelope addressed to my office. Please take a few minutes and record your perceptions of the work environment. Your current perceptions will serve as a benchmark as we attempt to improve the work environment by implementing Total Safety Management.

- **Time frame**

  Please return the completed form to my office no later than 4:00 p.m. on the 15th day of this month.

- **Confidentiality**

  Do not put your name on the survey form. Put the completed form in the envelope provided, seal the envelope, and drop it in the locked bin located outside my office. I will empty the bin at 4:00 p.m. on the 15th and summarize all responses. Individual forms will be destroyed. Only the summary sheet will be distributed.

- **Your Responses**

  Good or bad, we need your honest, objective input. Check the lines on the survey that most accurately reflect your *current* perceptions.

Thank you.

Figure 10–3
Measuring Employee Perceptions

TSM Facilitator typically conducts the focus group sessions, recording input on flip charts or marker boards. After each focus group session, the facilitator summarizes all input that represented the consensus of the group. Figure 10–4 contains several rules of thumb for successfully facilitating focus group sessions.

The internal focus group has several advantages, the most important of which are as follows:

- Inexpensive
- Allows the organization to collect in-depth input
- Interaction among participants can improve the quality and quantity of input

Figure 10–4

Rules of Thumb for Facilitators of Focus Groups

- Encourage participants to participate without dominating.
- Encourage participants to be open to diverse viewpoints.
- Encourage participants to *listen.*
- Do not become defensive and encourage participants to follow suit.
- Promote interaction among participants.
- Hold participants to the agenda; gently cut off tangents and wandering conversation.
- Keep the conversation focused without limiting reasonable debate.
- Summarize major points as the discussion progresses.
- Draw out nonparticipants.
- Restate for clarification points made by participants.
- Monitor reactions and bring to the surface any underlying issues.

The internal focus group also has disadvantages, the most important of which are as follows:

- Requires a high level of trust on the part of employees. If focus group members don't trust management and the facilitator, their participation will be both guarded and limited.
- Except in the smallest of organizations, only a representative sample of the overall workforce can participate.
- Time consuming

Survey (External)

This approach is the same as the internal survey with one exception: the survey is conducted and summarized by an outside agent. Using an external consultant can relieve the anxiety of employees about confidentiality. It can also solve the time problem experienced by large companies when summarizing survey input. The external survey, like the internal, involves all employees and can be administered in a relatively short period of time. Also like the internal survey, this approach limits the depth of responses employees are able to give, and it can be expensive.

Focus Group (External)

This approach is conducted just as the internal focus group is, except the focus group sessions now are conducted by an external consultant. Using an external consultant can give the employees a greater sense of security about confidentiality. Employees who don't yet have a high level of trust toward management may be more willing to open up to an external consultant.

This approach can be expensive, and it requires still more time than a survey. Although management is relieved of the time-consuming tasks of conducting focus group sessions and summarizing input, employees who participate in focus group sessions are still taken away from their jobs. Some organizations solve this problem by having the sessions conducted outside of working hours and paying employees who participate. Of course, this solution adds to the cost of the external focus group approach.

Regardless of the method used to measure initial employee perceptions, the data collected are summarized and presented to the TSM Steering Committee. These initial perceptions become the benchmark against which future perceptions are compared to determine if progress has been made.

═══════════ TSM CASE STUDY ═══════════

What Would You Do?

Mike Larson, TSM Facilitator for Atlantic Power Company (APC) is not sure what to do. He needs to benchmark employee perceptions concerning the current condition of APC's work environment. On the one hand, APC is small enough that he could conduct a survey or several focus groups himself. On the other hand, APC's employees don't seem to have a high level of trust in management. Larson is concerned about the quality and level of input he might get from employees. If you were Mike Larson, what would you do?

USING EMPLOYEE FEEDBACK

Regardless of whether employee feedback is collected using a survey or focus groups, it should be recorded in a quantifiable format such as the one shown in Figure 10–2. In this figure, each possible response can be assigned a numeric value as in the following example:

Strongly Disagree 1

Disagree 2

Agree 3

Strongly Agree 4

This allows an organization-wide average to be reported for each criterion measured and an overall score for the organization to be computed. Then, when employee perceptions are measured in the future, comparisons are easy to make. At such a time, progress—or the lack of it—is readily apparent for each criterion, as well as overall.

Figure 10–5 is a summary of an initial Employee Perceptions Survey for a manufacturing company. There are fifteen criteria in the survey document. The highest score

Figure 10–5
Sample Summary Sheet for
Employee Perceptions Survey

Employee Perceptions Tracking Chart	
Criteria	**Initial**
Fire Hazards	1.2
Electrical Hazards	1.3
Ergonomic Hazards	1.8
Bloodborne Hazards	1.9
Stress Hazards	2.0
Lifting Hazards	2.0
Automation Hazards	2.2
Mechanical Hazards	2.3
Noise/Vibration Hazards	2.4
Heat/Temperature Hazards	2.8
Impact Hazards	2.9
Toxic Hazards	3.1
Violence Hazards	3.2
Explosive Hazards	3.4
TOTALS	**32.5**

possible for an individual criterion is a 4. Consequently, the highest overall score possible is a 60.

To put to use the information shown in Figure 10–5, the TSM Steering Committee for a company would begin by prioritizing the criteria based on the average score for each. Lowest scores get the highest priority since they represent the most hazardous conditions. Based on the average scores, the criteria in Figure 10–5 would be prioritized as shown in the figure, with Fire Hazards shown to be the most pressing concern (a score of only 1.2), and Explosive Hazards the least pressing (score of 3.4).

With employee perceptions measured, summarized, and prioritized, and with the organization's safety and health weaknesses identified (Step 8), the TSM Steering Committee now has the information it needs to identify specific improvement projects. The perceptions summary and the list of organizational weaknesses form the basis for project selection in Step 12 of the implementation process.

REPEATING THE PROCESS

The point of benchmarking initial employee perceptions concerning workplace safety and health is to establish a base for gauging progress. Improvements will be made in an effort to create a safer, healthier work environment. Managers need to know if improve-

---

TSM TIP

Heinrich's Theory of Accident Causation

According to Heinrich, accidents are caused by predictable central factors such as human fault, unsafe conditions, or mechanical hazards. Removing the central factors prevents accidents and injuries.[2]

---

ments made have had the intended effect. This is the *check* phase of the *Assessment—Plan—Do—Check—Adjust* cycle. The check phase is accomplished by periodically repeating the employee perceptions survey.

Figure 10–6 shows a portion of a tracking chart that contains the results of successive employee perception surveys. In the initial survey, the organization's overall score was 32.5. Improvements were made in the areas indicated by the criteria, and another survey was conducted. This survey showed a marked improvement in employee percep-

	Employee Perceptions Tracking Chart			
Criteria	**Initial**	**2nd**	**3rd**	**4th**
Fire Hazards	1.2	3.1	3.3	
Electrical Hazards	1.3	2.9	3.1	
Ergonomic Hazards	1.8	3.4	3.4	
Bloodborne Hazards	1.9	3.1	3.1	
Stress Hazards	2.0	2.5	2.6	
Lifting Hazards	2.0	2.9	2.9	
Automation Hazards	2.2	3.1	3.3	
Mechanical Hazards	2.3	3.2	3.3	
Noise/Vibration Hazards	2.4	3.3	3.4	
Heat/Temperature Hazards	2.8	2.9	3.1	
Impact Hazards	2.9	3.2	3.3	
Toxic Hazards	3.1	3.8	3.8	
Violence Hazards	3.2	3.4	3.5	
Explosive Hazards	3.4	3.9	3.9	
TOTALS	32.5	44.7	46.0	

Figure 10–6
Results of Follow-Up Surveys

TSM TIP

Workers' Compensation Benefits

Workers' compensation benefits accrue to workers hurt on the job and to the families and dependents of employees who are fatally injured on the job. Typically the employee's spouse receives benefits for life or until remarriage. Dependents typically receive benefits until they reach the legal age of maturity.

tions (overall score of 44.7). Improvement efforts continued and the overall perception score increased again, but only slightly this time.

Such results are normal. When an initial survey is conducted working conditions are typically at their worst. When IPTs are activated, conditions begin to improve. The most noticeable improvements are often the initial improvements. Consequently, it is common to see overall perception scores level off somewhat after the second survey. This is only natural and should be expected. The *home runs,* so-to-speak, are typically hit early in the game because after the initial survey the organization is measuring improvements made to improvements.

It is important to ensure that managers and employees involved in the implementation of TSM don't become discouraged when employee-perception numbers begin to level off somewhat. Continuous incremental improvements over time are just as important as the more noticeable leaps forward made early in the process.

The overall score is just one indicator to observe. It is also important to compare scores over time on each individual criterion. For example, in Figure 10–6 the organization apparently made excellent progress in the area of fire hazards. The initial survey produced a score of 1.2 for employee perceptions of fire hazards. By the time the second survey was conducted, employee perceptions had improved markedly (score of 3.1). At the same time comparatively little progress was made in the area of heat/temperature hazards. The second score (2.9) is only slightly better than the initial score (2.8). However, the TSM Steering Committee apparently noticed the lack of progress and took appropriate action because the third survey shows a significant increase (3.1).

Improvements continue forever, as do employee perception surveys. Other indicators such as the number of days lost due to accidents, workers' compensation costs, employee absenteeism, and medical costs should also be analyzed every time an employee perceptions survey is conducted. All of these data, taken together, will give the TSM Steering Committee an accurate picture of how much real and perceived progress has been made. Both measures are important.

=== **STEP 10 IN ACTION** ===

Looking back, Mack Parmentier was pleased with the process he had used to benchmark employee perceptions at MPC. He had developed a survey document himself, but had convinced the TSM Steering Committee to contract with an outside consultant to conduct the survey and summarize employee feedback.

In order to clarify some of the summary input, Parmentier had then conducted several focus groups. At first, participants had been reluctant to speak out. However, Parmentier had patiently worked with them until, finally, the dam had broken and the input had begun to flow. The information provided in focus group sessions had added invaluable depth to that already collected by the survey. Consequently, the summary of input Parmentier presented to the Steering Committee was both thorough and comprehensive.

The executives on the TSM Steering Committee had been taken aback by the summary. Their collective response had been, "It can't be this bad, can it?" It had taken Parmentier some time and effort to convince the executives not to be offended by employee perceptions, but instead to accept them as a starting point. However, they had eventually conceded the point and switched to a *how-can-we-do-better* mode. Parmentier feels comfortable that the Steering Committee is now ready to tailor the implementation specifically for MPC.

SUMMARY

1. The rationale for benchmarking employee perceptions is to have a starting point to measure from. Without such a starting point, it is impossible to accurately and adequately measure progress.

2. There are four widely used methods for assessing employee perceptions. These methods are the internal survey, internal focus group, external survey, and external focus groups.

3. Employee feedback is collected in a way that allows it to be quantified. Each feedback criterion is assigned a numeric value. Then, in subsequent employee perception surveys, these numeric values are compared as a way to measure progress.

4. The process of quantifying employee perceptions is repeated periodically forever. This is how progress is measured. The timing of each successive repetition is important. Enough time has to elapse between employee perception surveys to allow improvements to occur. A good rule of thumb is to have at least six months between surveys.

KEY TERMS AND CONCEPTS

Benchmarking	Focus group (internal)
Employee perceptions survey	Survey (external)
Focus group (external)	Survey (internal)

REVIEW QUESTIONS

1. Explain briefly the rationale for assessing employee perceptions concerning the state of the work environment.

2. Describe the various methods that are widely used for assessing employee perceptions.

3. How is employee-perception information used once it has been collected?

4. Why is the assessment of employee perceptions repeated periodically? How often should the process be repeated?

ENDNOTES

1. National Safety Council, *Accident Facts* (Chicago: 1995) p. 40.

2. David L. Goetsch, *Occupational Safety and Health in the Age of High Technology*, 2nd ed. (Upper Saddle River, N. J.: Prentice Hall, 1996), p. 35.

Tailoring the Implementation to the Organization

At this point the TSM Steering Committee knows the organization's strengths and weaknesses, likely advocates, resisters, and employer-perceptions concerning the state of the work environment. In Step 11, this information is used to tailor the remainder of the TSM implementation process for the specific organization in question.

RATIONALE FOR TAILORING THE IMPLEMENTATION

Although the basic steps in the implementation process are the same for all organizations, what actually occurs within each step differs from organization to organization. This is because no two organizations have the same strengths and weaknesses, nor the same workplace conditions. In addition, there is a predictable list of potential roadblocks to successful implementation. Figure 11–1 is a list of these potential roadblocks. Which of these roadblocks actually exist, and the extent to which they exist can differ markedly from organization to organization.

The rationale for tailoring the implementation is that success is more likely if the specific roadblocks in a given organization are identified and adequately addressed by the tailoring plan. The roadblocks that organizations are most likely to face in varying degrees are explained in the following subsections.

Figure 11–1
Roadblocks to Implementing
TSM

**Checklist of
Potential Roadblocks to Implementing TSM**

✓ Lack of leadership
✓ Lack of top-management support
✓ Lack of buy-in from middle management
✓ No definite course of action
✓ Vacillation
✓ Conflicting agendas
✓ Failure to exploit strengths
✓ Failure to address weaknesses
✓ Failure to commit sufficient time
✓ Change-resistant culture
✓ Insufficient understanding of TSM

Lack of Leadership and Support

Change of any kind can be difficult in an organization. Without leadership and support from top management it can be impossible. The organization would not have gotten to this point (Step 11) without leadership and support. The key now is to ensure that leadership and support continue. Any vacillation on the part of top management can bring the implementation to a screeching halt. Leadership and support in this step manifest themselves in the effort which members of the TSM Steering Committee put into tailoring the implementation. Making sure that Steering Committee members understand this and stay committed is the responsibility of the TSM Facilitator.

Lack of Buy-In from Middle Management

TSM is a concept that involves employees in the decision making process and empowers them to offer input and feedback continually. Empowerment does not mean that management abdicates its responsibility for making decisions. Managers are still responsible for making decisions relating to safety and health issues. By empowering employees, managers engage them in a way that leads to better, more informed decisions.

Empowerment does mean that employees *are* able to make decisions within clearly prescribed limits, limits that are established by management. Both aspects of empowerment run counter to the traditional *management-thinks-and employees-do* philosophy. As a result, middle managers who subscribe to this philosophy often reject TSM—at least in the beginning. Consequently, it is important to consider the extent of middle management buy-in when tailoring the remainder of the implementation process.

Middle managers who are likely to be resisters have already been identified. To this point they have been excluded from the implementation process. In tailoring the remainder of the implementation, middle managers who are still considered likely

---

TSM TIP

Hazard Analysis

IPT members will need to understand hazard analysis as a process. Hazard analysis is a systematic process for identifying hazards and for recommending corrective action. IPT members will be concerned about the following factors: (a) the likelihood of an accident or injury being caused by the hazard, and (b) the severity of injury, illness, or property damages that could result from the hazard.

---

resisters should continue to be excluded. This will decrease the likelihood of roadblocks that might keep TSM from gaining a foothold.

No Definite Course of Action

The remainder of the implementation must be accurately focused. There must be a definite course of action that is thoroughly understood and strictly adhered to by all members of the TSM Steering Committee and all other employees involved in the implementation. This definite course of action is established by the plan developed in this step for tailoring the remainder of the implementation. This plan will ensure that the entire organization speaks with one voice.

Vacillation

Once the TSM Steering Committee has developed a plan for the remainder of the implementation, each member should set a positive example of carrying out the plan with determination and purpose. At this point in the process doubters and resisters still have the advantage. Vacillation on the part of executive management will only enhance that advantage by giving credibility to their doubt.

Sticking to the plan does not mean forging ahead with blind stubbornness if the plan clearly isn't working. Making adjustments to get around unexpected barriers is not vacillation. Remember the A-P-D-C-A Cycle. Any time a plan is implemented, results are checked and adjustments are made as needed. Vacillation manifests itself at this point as reluctance to implement the plan.

Conflicting Agendas

Even at this state in the process, conflicting agendas among members of the TSM Steering Committee might surface. It is important that conflicting agendas be handled with a high degree of finality in this step. Beyond this step they are likely to sabotage the positive results the organization needs to see from TSM. This is the responsibility of the organization's CEO. It is a good idea at this point for the CEO to review the organization's safety and health vision, guiding principles, and objectives. A frank and open discussion with all members of the Steering Committee using the safety and health plan as the focal point is the recommended approach for determining if all members are working toward the same end.

Failure to Exploit Strengths and Overcome Weaknesses

In Step 8, The TSM Steering Committee identified the organization's strengths and weaknesses from the perspective of safety and health. Now this information is put to active use. In developing a plan to tailor the remainder of the implementation, the TSM Steering Committee will consider the organization's strengths and weaknesses, the list of likely advocates and resisters, and the summary of employee perceptions. This is the step in which all of this information is put to use.

Failure to Commit Sufficient Time

Patience is an invaluable, albeit rare virtue. Organizations tend to be just like people in that once a course is set, they want to see results immediately. This fact, more than any other, accounts for the success of fast food restaurants, microwave ovens, express mail, and a long list of *instant* products (coffee, oatmeal, pudding, etc.). Unfortunately, organizational change is neither instant or easy. Rather, it requires a concentrated effort exerted over time.

In the present step, members of the Steering Committee should recommit themselves to giving TSM the time it needs to produce results. The amount of time will vary from organization to organization depending on the state of the work environment before the implementation and the effectiveness of the implementation. An organization that starts out with a deplorable work environment is likely to see noticeable results faster than one that starts out with just moderate safety and health problems; this is so provided, of course, the implementation process is carried out effectively. The key is to apply the A-P-D-C-A Cycle, adjusting as necessary, and then to give the process time to work.

Change-Resistant Culture

An organizational culture is by its very nature resistant to change. Organizational cultures develop slowly over time, and once established, become self-perpetuating. Consequently, in this step executives should remind themselves, each other, and all employees involved in the implementation to be persistent in addition to being patient. Trying to change an organizational culture is like trying to change the course of a stream. The task will clearly be difficult. But if one keeps working patiently and persistently for long enough, it can be done.

Insufficient Understanding of TSM

This problem is most likely to surface in Step 13 when IPTs are actually chartered and activated. Before employees serve on an IPT, they will be trained. This training must equip them with a thorough understanding of both the *why* and the *how* of TSM.

The roadblocks described herein may or may not be present in an organization that is implementing TSM. The key is to determine which roadblocks are actually present and to deal with them in the plan for tailoring the remainder of the implementation.

PLANNING THE REMAINDER OF THE IMPLEMENTATION

This step must be planned and tailored, giving special attention to the following factors (Figure 11–2).

■ Likely advocates and resisters
■ Organizational strengths and weaknesses
■ Current employee perceptions concerning the work environment

In the next step (Step 12), the organization will select the initial projects in an improvement process that will continue forever.

Part 1: Advocates and Resisters

Likely advocates and resisters were identified in Step 9. In this step the TSM Steering Committee reviews the list one last time and makes adjustments as necessary. Between Steps 9 and 11, the Steering Committee's perception of one or more employees may have changed. An individual who was originally categorized as a likely resister may now be viewed as a likely advocate or vice-versa. Such reversals are common. This is why it is important to give the advocate-and-resister list one last look.

Once the list has been updated, potential advocates are entered in Part One of the plan by department/unit and position, as shown in Figure 11–3. Advocates are listed as "employees chosen to participate initially in the TSM implementation." Potential resisters are simply left off the list. The plan developed in this step is not intended for

Figure 11–2
Factors to Consider in
Developing the Plan for Tailoring
the Remainder of the
Implementation

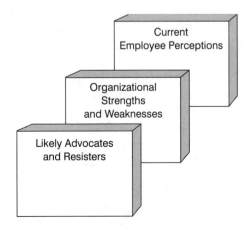

Figure 11–3
Partial List of Likely Advocates
by Department

Part 1: Advocates by Department

Accounting

Davis, Jane . . . .   Bookkeeper
Crawford, Faye . . . .   Accounts Receivable
Smith, Alex . . . .   Accounts Payable

Engineering

Durham, Allen . . . .   Industrial Engineer
Evans, Myra . . . .   CAD Technician
Edwards, Wanda . . . .   Mechanical Engineer
Freemont, Marquis . . . .   Electrical Engineer
Ganmer, Barr . . . .   CAD Technician
Lewis, Dan . . . .   Industrial Engineer
Norman, Andrea . . . .   Electrical Engineer
Roberts, Lamar . . . .   Mechanical Engineer
Williamson, George . . . .   CAD Technician

Manufacturing

Dante, Mary . . . .   Painting Technician
Donalson, Mike . . . .   Machine Operator
Douglas, Nick . . . .   Machinist
Ortega, Guadalupe . . . .   Assembler

general circulation. It is developed by the Steering Committee for the use of its members. However, it is not a secret document. In fact, employees should be welcome to review it. Some organizations make copies available in a central location that is convenient and accessible for all employees.

Since the contents of the plan will and should be widely known, it is important that it be free of any indicators that might be misinterpreted by employees. Employees will accept that only a limited number of them can be involved initially. They will not, however, be comfortable with categorizations that might, at least on the surface, appear to threaten their status and job security. Remember, in time, initial resisters can become staunch advocates.

Part 2: Organizational Strengths and Weaknesses

In Step 8, the organization's strengths and weaknesses in the area of health and safety were identified. Those strengths and weaknesses are now entered in Part 2 of the plan.

TSM TIP

Accident Investigations

Accident investigations are used in a TSM setting to collect facts concerning cause, not fault. Fault-finding will serve only to make those with knowledge of the accident reticent to share what they know. Accident investigations should be structured to determine who, what, when, where, and why.

Figure 11–4 is an example from Part 2 of the plan for a manufacturing organization. The tailoring aspect of this component involves answering the following questions:

- How can we use our strengths to promote and facilitate the remainder of the implementation?
- Which of our weaknesses must be dealt with by the Steering Committee?
- Which of our weaknesses should be turned over to an IPT?

Referring to Figure 11–4, consider how the TSM Steering Committee for that organization might answer these questions. The first question has to do with taking advantage of the organization's strengths to promote a successful implementation. All three of the strengths will promote a successful implementation.

Commitment on the part of top-level managers is essential. In fact, the implementation would not have even gotten started without it. The fact that the organization's employees are interested in seeing the work environment improved is also a plus. With such a strong mutual commitment in place, the remainder of the implementation should proceed at a steady pace. However, the strength that might best be used to really get the ball rolling so to speak is the presence of a safety-minded supervisor in the organization's Hazardous Waste Management Department. If the initial IPT is chaired by the supervisor in question, chances of a successful initial project are high. Involving the subject supervisor in this way might be the answer to the question of how to take full advantage of the organization's strengths to promote a successful implementation.

As to which of the organization's weaknesses must be dealt with by the TSM Steering Committee, there are two considerations. First, is there a weakness so pressing that it must be dealt with immediately? Second, are there weaknesses that, because of their nature, only executive management can handle? Taking these considerations into account, examine the list of weaknesses in Figure 11–4. Only the first of them—no organizationwide safety and health program—would need to be handled at once by executive management, by the implementation of TSM. All of the other weaknesses can be referred to one or more IPTs for study and recommendations. This fact answers the final question for Part 2. Answering these questions is the tailoring aspect for Part 2.

Part 2: Strengths and Weaknesses

Strengths

- Commitment of top management to the full implementation of TSM
- Safety-minded supervisor of the Hazardous Waste Management Department
- Strong interest of employees in improving the company's work environment

Weaknesses

- No well-coordinated organization-wide safety and health program
- Few established safety/health procedures
- Little or no use of personal protective equipment
- Insufficient knowledge of safety and health issues/concerns on the part of employees at all levels
- High accident/injury rate with attendant results (e.g., lost days due to accidents, high medical costs, high absentee rate)
- Level of ergonomic hazards (high)
- Level of noise/vibration hazards (high)
- Level of stress hazards (moderate)
- Level of electrical hazards (moderate)

Figure 11–4
Sample List of an Organization's Strengths and Weaknesses

Part 3: Employee Perceptions

Employee perceptions concerning the current state of the workplace were identified in Step 10. These perceptions are now recorded in Part 3 of the plan for the organization we are studying (see Figure 11–5). It contains six areas of concern. The hazard level in the first four areas is considered high by employees. The other two have moderate hazard levels. The tailoring aspect of Part 3 involves prioritizing the problems listed therein.

The organization is not yet mature enough with regard to TSM to take on simultaneously the problems listed in Part 3. Eventually there will be IPTs working on all of these problems. However, until the organization has some experience in applying the TSM philosophy, it is best to be selective in the assignment of projects to IPTs.

The guiding principle in prioritizing the problems in Part 3 is timing. In deciding which of the problems must be dealt with first, second, third, and so on, the Steering Committee should consider the following criteria:

- The perceived hazard level of the problem (high, moderate, low)
- Relative potential for immediate danger of the problem
- Documented safety and health problems relating to the problem

Figure 11–5

Example Summary of Employee
Perceptions

Part 3: Employee Perceptions

The hazard levels currently present in the
work environment are viewed by employees
as follows:

Hazard	Level
Lifting Hazards	High
Mechanical Hazards	High
Ergonomic Hazards	High
Noise/Vibration Hazards	High
Stress Hazards	Moderate
Electrical Hazards	Moderate

Consider how these criteria might be applied to the list in Figure 11–5. The perceived hazard level of each problem is already rated (high, moderate). These levels were derived by converting the numeric values from the employee-survey summary into qualitative descriptors (e.g., 0-2 = high hazard level; 3-4 = moderate level; 5-6 = low hazard level).

Of the four high-hazard areas, one of them—namely, ergonomic hazards—has less immediacy than the others. This is because ergonomic injuries are typically the result of repetition over time. However, the other high-hazard areas—to wit, lifting, mechanical, and noise/vibration—can cause injuries immediately. Consequently, it is necessary to go further and consider the third criteria, documentation, before prioritizing these three problems.

By examining the organization's accident records the TSM Facilitator can determine which of the high-hazard factors is associated with the most accidents, the most serious accidents, and the most costly accidents. This exercise does more than just produce the information needed to prioritize problems, it also gives the TSM Facilitator one last chance to compare factual records with employee perceptions. This was done in Step 10, but one last look is always a good idea.

TSM TIP

Safety Rules and Regulations

From a legal standpoint, an organization's obligations regarding safety and health rules and regulations are as follows: it must

■ *Have rules and regulations that ensure a safe and healthy workplace.*

■ *Ensure that all employees are knowledgeable of the rules and regulations.*

■ *Ensure that rules and regulations are enforced objectively and consistently.*

INCORPORATING THE A-P-D-C-A CYCLE

In the next step, projects will be selected for assignment to IPTs. This step will begin a process that is ongoing forever. The model for this on-going process is the A-P-D-C-A Cycle, repeated because of its importance in Figure 11–6. The TSM Steering Committee will apply the model as the basis for operating a TSM organization. Members of the TSM Steering Committee have already been introduced to the A-P-D-C-A Cycle. Even so, it's a good idea to review the model in this step. The paragraphs that follow describe how the Steering Committee and an IPT apply the model to continually improve the quality of the work environment.

Assess Step

The TSM Steering Committee with the assistance of the TSM Facilitator, continues to periodically assess the state of the workplace. Assessment methods include the following: (Figure 11–7):

■ Organizational-weakness surveys
■ Employee-perception surveys

Figure 11–6
Applying the A-P-D-C-A Cycle,
Once Again

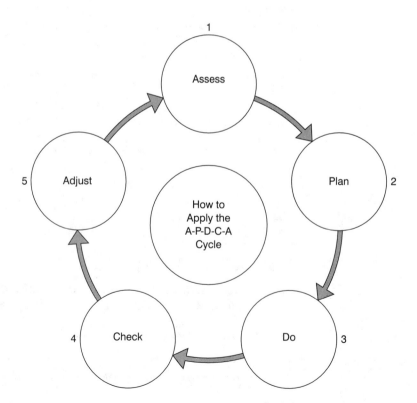

Figure 11–7
Assessment Methods Used
Forever

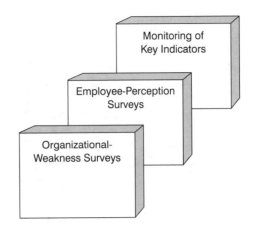

- Monitoring of key indicators (e.g., medical costs, workplace injuries by type, absenteeism, lost days due to accidents)

Organizational-weakness and employee-perception surveys are conducted periodically: at least once each year, but not more than twice. Key indicators are monitored continually by the TSM Facilitator.

Plan Step

The plan developed in Step 11 is a living, changing document. Based on the periodic reassessments made by the Steering Committee and Facilitator, the plan developed herein is updated periodically. As the TSM implementation progresses and improvements are made, things will change. Resisters will become advocates, organizational weaknesses will be corrected, employee perceptions will improve, and key indicators will change. These changes should be reflected in the plan.

Do Step

This is the step when the TSM Steering Committee either takes direct and immediate action to correct a safety and health-related problem or refers the problem to an IPT. As one problem is solved, others will surface. An improvement made today will need to be improved even further tomorrow. This is the continuous improvement aspect of TSM, and it never stops.

Check Step

As IPTs are chartered, they are given specific assignments that are undertaken within a specified time frame. Their progress is monitored by the TSM Facilitator, on behalf of the Steering Committee. Is the IPT staying on task? Is the IPT ahead of schedule, on schedule, or behind schedule? Is the IPT being inhibited by unexpected roadblocks?

Adjust Step

If an IPT falls behind schedule, gets off course, or encounters roadblocks, the TSM Facilitator—working on behalf of the Steering Committee—helps make the necessary adjustments. Necessary adjustments might be relatively minor. For example, the Facilitator might need to pull the IPT back from a tangent and guide it back on course. On the other hand, adjustments might be major. For example, the IPT may need funding to bring in an expert to study a problem and make recommendations. In any case, adjustments are made as necessary with the Facilitator serving as the link between IPTs and the TSM Steering Committee.

=== TSM CASE STUDY ===

What Would You Do?

Jack Morgan, Director of Safety and Health, was stumped. The TSM Steering Committee for Intercontinental Apparel Company (IAC) had just finished putting the final touches on its plan for completing the company's implementation of TSM. The plan was a good one, and Jack Morgan was understandably proud of it. But any celebration that might be in order would have to wait. Right now Morgan wasn't sure how to respond to one of the Steering Committee members who had just said, "Now that we have a plan, what do we do next? I know we are supposed to finish the remaining steps of implementation, but how TSM becomes permanent and ongoing isn't clear to me." If you were Jack Morgan, how would you respond?

=== STEP 11 IN ACTION ===

Knowing that having a well-thought-out plan for the remainder of the TSM implementation would be critical, Mack Parmentier had done his homework carefully. Also knowing that developing such a plan would be a new experience for the Steering Committee members, he had taken the time to develop a rough draft of a plan so as to have a *strawman* to work from.

Parmentier's draft plan had three components: (a) an alphabetized list of all of MPC's employees with each name followed by either an "A" for advocate or an "R" for resister; (b) a summary of the organization's strengths and weaknesses; and (c) a summary of current employee perceptions. In anticipation of his meeting with the Steering Committee, Parmentier had listed all of MPC's employees in Part 1, despite knowing that this is not the recommended approach. He had two reasons for doing this. First, he wanted to avoid preempting the Steering Committee. Developing the plan for the remainder of the implementation is the responsibility of the Steering Committee. Parmentier's rough draft was intended as an *icebreaker*, not a finished product. Second, he wanted to ensure that the Steering Committee went through the exercise of reviewing the list one more time before finalizing it.

In Part 2, Parmentier had listed the organization's strengths and weaknesses without comment or analysis. This approach was used because it would allow Steering Committee members to do the necessary analysis themselves. Parmentier wanted to ensure that the Steering Committee asked such questions as:

- How can we use our strengths to greatest advantage during the remainder of implementation?
- Which weaknesses should be dealt with directly by the Steering Committee and which should be assigned to IPTs?
- What safety and health problems should be pursued first, second, third, and so on?

In Part 3, Parmentier had simply copied the summary of current employee perceptions that had already been presented to the Steering Committee. He had reasoned that including the summary would give the Steering Committee one more opportunity to prioritize problems discovered during the employee-perceptions survey.

The Steering Committee's planning meeting had gone well. In looking back, Parmentier was now convinced that his strawman idea had been a good one. The draft plan had broken the ice and gotten the Steering Committee focused without preempting its responsibilities. The draft plan had been Parmentier's, but the final plan was clearly the work of the Steering Committee.

Once the finishing touches had been applied to the plan, Parmentier had reviewed the A-P-D-C-A Cycle with the Steering Committee. This part of the meeting had gone especially well. One of the Steering Committee members had made the comment, "At last I understand how it all fits together." Several members had agreed. Parmentier's final thought on Step 11 was simple. "We're getting there," he mused.

=========== SUMMARY ===========

1. No two organizations have the same strengths, weaknesses, key indicators, or safety/health problems. Consequently, implementation plans must be tailored for the specific organizations in question.

2. Problems commonly confronted during a TSM implementation are as follows: lack of leadership, lack of support, lack of buy-in from middle management, failure to exploit strengths, failure to overcome weaknesses, failure to commit sufficient time, no definite course of action, vacillation, conflicting agendas, change-resistant culture, insufficient understanding of TSM.

3. In planning the remainder of the implementation, three major factors should be considered: (a) likely advocates and resisters; (b) organizational strengths and weaknesses; and (c) current employee perceptions concerning the work environment.

4. Once the plan developed in this step is completed, the TSM Steering Committee should review the A-P-D-C-A Cycle. This cycle is the model that will be used to make TSM the permanent and ongoing approach to managing the safety and health of the work environment.

KEY TERMS AND CONCEPTS

A-P-D-C-A Cycle

Change-resistant culture

Conflicting agendas

Failure to commit sufficient time

Failure to exploit strengths

Failure to overcome weaknesses

Insufficient understanding of TSM

Lack of buy-in from middle management

Lack of leadership

Lack of support

No definite course of action

Vacillation

REVIEW QUESTIONS

1. What is the rationale for tailoring the implementation?

2. List and explain briefly the roadblocks that are commonly encountered when implementing TSM.

3. What factors should be considered when developing a tailored plan for the remainder of the implementation?

4. Explain why a TSM Steering Committee might choose to deal with a specific safety and health problem directly instead of referring it to an IPT.

5. Explain how the A-P-D-C-A Cycle is used for the remainder of the TSM implementation and beyond.

=== STEP TWELVE ========================

Identify Specific
Improvement Projects

=== MAJOR TOPICS ======================

■ Rationale for Selecting Initial Projects Carefully
■ Criteria for Project Selection
■ Defining the Project's Content and Scope
■ Deciding How Many Projects to Implement Initially
■ Step 12 in Action

In Step 11, the TSM Steering Committee developed a plan for tailoring the remainder of the implementation. In this step, the Steering Committee decides which specific project or projects in the plan will be given to IPTs first. These projects are clearly defined in terms of both scope and content before they are turned over to an IPT in the next step.

RATIONALE FOR SELECTING PROJECTS CAREFULLY

Step 12 is the last in the Identification and Assessment phase of the TSM implementation, and it is a critical step. The care with which the first project to be assigned to an IPT is selected can determine whether the TSM implementation ultimately succeeds or fails. Careful project selection in and of itself will not guarantee a successful implementation. But haphazard selection can guarantee failure. Even at Step 12 of a 15-step implementation process, the concept is still too new and too fragile to be subjected to a major setback.

Although some organizations have failed with their first improvement project and still gone on to successfully implement TSM, this is not the norm. The more common result is when the first improvement project fails, the implementation sputters and the organization has to work doubly hard to get back on track. On the other hand, success breeds success. Consequently, the Steering Committee should consider all applicable criteria carefully before selecting the first project.

CRITERIA FOR PROJECT SELECTION

Most of the criteria for selecting projects were applied in Step 11 when the tailored plan was developed. Figure 12–1 is a document that was excerpted from the tailored plan developed in Step 11. This is the type of excerpt the Steering Committee should use in selecting initial projects. It contains the following information: (a) the organizational strength that has been selected by the Steering Committee as having the most potential for getting the remainder of the implementation off to a good start; (b) the organizational weaknesses that are to be referred to IPTs; and (c) the highest priority safety and health problems as identified in the employee-perceptions survey and verified by an examination of accident reports.

Figure 12–2 contains the final criteria for selecting an initial project(s). The criteria are as follows: the project

- Should take advantage of at least one organizational strength.
- Should attack one of the organization's most serious weaknesses.
- Should attack one of the organization's most serious safety problems.
- Should have a good chance of succeeding.

Takes Advantage of One Organizational Strength

It is critical that the first project assigned to an IPT succeed. This will get the final stage of the implementation off to a good start, and success will breed success. It will also begin to silence critics, knock fence sitters off the fence, and convert resisters. One of the best ways to ensure an initial success is to take advantage of one or more of the organization's strengths in the area of safety and health.

The excerpt in Figure 12–1 contains the following organizational strength: "*Safety-minded supervisor in the Hazardous Waste Management Department.*" By including this

Figure 12–1
Excerpt from Implementation Plan

Most Beneficial Strength (Initially)

Safety-minded supervisor in the Hazardous Waste Management Department

Most Serious Weakness (Initially)

High accident/injury rate resulting in high medical costs, and an unacceptable level of absenteeism

Most Serious Safety Problems (Currently)

1. Lifting hazards
2. Mechanical hazards
3. Noise/vibration hazards

Figure 12–2
Final Criteria for Initial Project
Selection

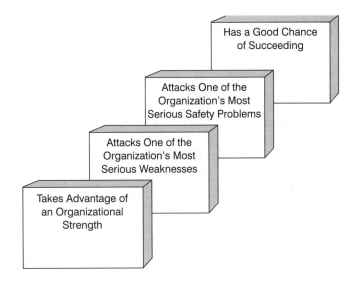

particular strength in the excerpt from its tailored plan, the TSM Steering Committee under discussion has made a judgement call. In the judgement of its members, this is the organizational strength that has the most potential for ensuring the success of the first project.

The way to take advantage of this strength is to appoint the supervisor in question to the initial IPT—or, better yet, let him chair it. Being a supervisor, this individual will be accustomed to coordinating the work of a team. Being a safety/health advocate, he or she will want the IPT to succeed and, therefore, will probably be willing to work hard to ensure that it does.

Attacks One of the Organization's Most Serious Weaknesses

The organization's weaknesses from a safety and health perspective were identified in Step 8. These weaknesses were prioritized as part of the tailoring process in Step 11. In this step, the Steering Committee will ensure that the initial project will attack a high-priority weakness. This will give the project significance, and it will ensure that at least

---

TSM TIP

Electromagnetic Radiation Hazards

Electromagnetic radiation, or EMR, is a common form of nonionizing radiation. It emanates from various electrical sources including power lines, electrical appliances, radios, and televisions. The major concern about EMR is a suspected causal link to leukemia.[1] Modern safety and health professionals should monitor employee exposure to EMR carefully.

one important step is taken to improve the work environment in a noticeable way. On the one hand, the initial project should be one that has a good chance of succeeding. On the other hand, however, it cannot be perceived by employees as a *ho-hum* project.

For example, in Figure 12–1, the TSM Steering Committee has selected the following area of weakness as one needing immediate attention: "High accident/injury rate which has resulted in high medical costs and an unacceptable level of absenteeism." Certainly this is a weakness that must be dealt with soon.

Attacks One of the Organization's Most Serious Safety Problems

Specific safety and health problems were identified during the employee-perceptions survey (Step 10), and then verified by examinations of key indicators in the organization's safety and health records. Now one of the specific, high-priority problems must be selected for assignment to an IPT. Ideally, the one selected will be the problem that has the most immediate potential for causing an injury, incident, or damage to property. It should also be a problem that relates closely to the area of weakness that is to be attacked. In other words, the specific safety/health problem selected should be one that contributes in a significant way to the organization's highest-priority area of weakness.

In Figure 12–1 there are three specific safety/health problems listed in priority order. All three of these represent serious safety/health problems, and all three relate directly to the organizational weakness selected earlier.

Has a Good Chance of Succeeding

The importance of successful early-stage projects has been explained at length already. One of the ways this issue is spoken to is by incorporating at least one of the organization's primary strengths when assigning initial projects to IPTs. This strategy was explained earlier.

Another strategy that will promote success is to select an initial project that can be dealt with using existing resources and within a relatively short period of time. There is nothing to be gained from selecting a project that will require more resources (money, personnel, etc.) than the organization can commit within a reasonable time frame.

For example, say the organization must choose between two potential initial projects. Both meet the criteria set forth up to this point. One project has the potential to

---

TSM TIP

Noise Hazards

The principal hazard associated with excessive noise is hearing loss. Exposure to excessive noise levels for an extended period of time can damage the inner ear to the extent that the ability to hear high-frequency sound is diminished or even lost altogether.[2]

take as much as six months and could require the expenditure of thousands of dollars from a tight budget that has already been set. The other project can probably be completed in a matter of weeks with only minimal expenditures during the current quarter of the fiscal year. The latter project probably makes more sense as an initial project.

DEFINING THE PROJECT'S CONTENT AND SCOPE

By applying all applicable criteria, the TSM Steering Committee selects an initial project. Before this project is assigned to an IPT it must be clearly defined in terms of scope and content. This definition of scope and content becomes the charter for the IPT to which the project is assigned.

Assume that the initial project selected by the Steering Committee is *lifting hazards* (from Figure 12–1). This project relates directly to the organizational weakness contained in Figure 12–1 (high-accident/injury rate). The Steering Committee now develops a charter that will guide the IPT to which this project is assigned.

Figure 12–3 is an example of a charter that was developed for the lifting-hazards project. The *background* component describes the area of weakness that is being

Figure 12–3
Sample IPT Charter

Charter
Lifting Hazards Project

Background

One of our most pressing weaknesses in the area of safety/health is the organization's high accident/injury rate. This weakness results in both high medical costs and an unacceptable level of absenteeism. One of the safety/health problems that contributes directly to this weakness consists of *lifting hazards.*

Project Definitions/Purpose

The purpose of this project is to identify specific lifting hazards in the workplace and to make recommendations for either eliminating or overcoming those hazards.

Project Goals

1. Identify lifting hazards in the workplace.
2. Prioritize the hazards from most critical to least critical.
3. Make recommendations to the Steering Committee (through the TSM Facilitator) for eliminating/overcoming the hazards.

Time frame

Complete the project in four weeks from initiation date.

attacked, and the specific project within that area. The *project definition* component explains the IPT's purpose. The purpose in this case has two parts: (a) identification of lifting hazards; and (b) submission of recommendations for eliminating or overcoming the lifting hazards identified.

The *project goals* component gives three specific goals the IPT will be expected to accomplish. These three goals exemplify the kind that are typically assigned to IPTs. The formula statement *identify-prioritize-recommend* is a good summary of what most IPTs actually do.

The *time frame* component gives the IPT the information it will need to establish a work schedule. Unless otherwise specified, the time frame should be viewed as a guideline rather than a hard-and-fast deadline. It is the TSM Facilitator's best estimate of how long it should take a group of employees who can break away from their jobs for about an hour a day to accomplish the goals set forth in the IPT's charter.

It might take the IPT less than the allotted time to complete its tasks, or it might take more. It is the TSM Facilitator's job to work closely with the IPT's chair, to monitor progress, and to make judgements about the performance of the team. If the team fulfills its charter in a relatively short period of time, the Facilitator must decide if an appropriate amount of effort has been put forth, or if the IPT has conducted only a cursory investigation of the problem. If the team is taking longer than the allotted time to fulfill its charter, the Facilitator must decide if the IPT has dropped the ball, or legitimately needs more time.

Productive interaction between the TSM Facilitator and the IPTs will be a critical factor in determining the on-going, long-term success of TSM. This interaction, coupled with the Facilitator's work with the Steering Committee will be the *grease* that lubricates the wheels of the TSM machine forever.

DETERMINING HOW MANY PROJECTS TO IMPLEMENT INITIALLY

How many projects to begin with is always an open question, and the answer can vary from organization to organization. Without getting into actual numbers just yet, a good rule of thumb to apply in determining how many projects to implement initially is *slow but steady is best*. With any new concept in which change is implicit, it is better to start slowly and progress deliberately than to take on more than can be properly handled.

Having gone through all of the preliminary stress, it is only natural that members of the TSM Steering Committee will be anxious to get teams working on problems. However, at this point the Steering Committee has no experience in applying the A-P-D-C-A Cycle in the coordination of even one team. In addition, at this point everything is new to employees assigned to IPTs. Consequently, to activate more than a few teams initially is ill-advised, and can lead to chaos instead of progress.

With the *slow and steady* rule of thumb in mind, the discussion can now move to actual numbers. If the implementation has progressed smoothly up to this point and Steering Committee members have a high level of confidence in themselves, two or three teams might be activated initially. If the level of confidence expressed by Steering

Committee members is not high, begin with just one team. Then, as confidence grows and experience is gained, the number of teams working simultaneously can be increased accordingly.

TSM CASE STUDY

What Would You Do?

Should he recommend starting with just one improvement project, or should he propose that the Steering Committee initiate several at one time? John Brown wasn't sure how to answer this question at the moment. "No need to panic," thought Brown. "I still have two days before my recommendation is due to the Steering Committee. I'll call a couple of colleagues and discuss the issue with them." If you were one of the colleagues John Brown called, what would you tell him?"

STEP 12 IN ACTION

Mack Parmentier held in his hand the charter for MPC's *first improvement project*. He had expected that developing the initial charter would be a challenge, and it had been. But the greater challenge had been to convince the Steering Committee to begin with just one project. Several committee members who had more enthusiasm than expertise had pushed hard for chartering five to ten teams simultaneously. After working so hard to generate this level of enthusiasm, Parmentier wanted to do nothing to dampen it. Consequently, he had found himself facing an interesting dilemma.

At this point in the implementation process, "Go slowly" was not what the Steering Committee wanted to hear. The Committee was like a football team that had gone through all of the preliminary activities and build-up necessary to get ready for the Super Bowl. Then, just before the long anticipated kickoff, the coach says, "Slow down." As reluctant as he was to reign them in, Parmentier didn't want MPC's Steering Committee to be like the football team that becomes so anxious to run the ball that it gets careless and fumbles.

He knew the Steering Committee needed to go slowly until it gained experience. He knew mistakes would be made and that with several new teams working simultaneously, those mistakes would be multiplied. The argument that eventually won the Steering Committee over to Parmentier's way of thinking had had nothing to do with safety.

Reasoning that as executives in a processing company, the Steering Committee members would relate best to a process-oriented argument, Parmentier had used one. He had posed the question, "Gentlemen, when you design and build a new processing unit, do you simply open the pipes and start processing chemicals, or do you run the process through initial tests to work out the bugs?" All present had answered as one, "We work out the bugs first, of course." Parmentier had argued that initiating the first IPT was similar to initiating a new process, and that even the most successful first projects encounter bugs that have to be worked out.

The logic in Parmentier's argument had swayed even the most enthusiastic committee members, but even so he knew it would be wise to get an IPT working right away. Consequently, Parmentier had gotten the Steering Committee focused on developing a charter that attacked the problem of chemical burns and related injuries. This problem could be linked directly to one of MPC's most serious weaknesses; high medical costs. It would also take advantage of one of the company's most beneficial strengths; Parmentier's personal experience and expertise in chemical-oriented accident prevention.

In reviewing the charter that had been developed, Parmentier felt confident that MPC's first improvement project would be a success. He looked forward to getting an IPT formed and working.

SUMMARY

1. Careful project selection in and of itself will not guarantee a successful implementation, but haphazard selection can guarantee failure. This is the rationale for selecting initial projects carefully.

2. The final criteria for selecting initial projects are that they should take advantage of at least one organizational strength; should attack one of the organization's most serious weaknesses; should attack one of the organization's most serious safety problems; and should have a good chance of succeeding.

3. An improvement project's content and scope are defined in a charter. The charter consists of the following: (a) a background statement, (b) a statement of purpose (project definition); (c) project goals, and (d) a time frame.

4. The number of projects implemented initially depends on the ability of the Steering Committee to apply the A-P-D-C-A Cycle as a monitoring tool. A Steering Committee that is confident in its ability at this point might begin with two or even three projects. A Steering Committee that is not confident should begin with just one project.

KEY TERMS AND CONCEPTS

A-P-D-C-A Cycle	Project definition
Background	Project goals
Charter	Selecting projects carefully
Good chance of succeeding	Slow and steady is best
Identify-prioritize-recommend	Time frame

REVIEW QUESTIONS

1. Explain the rationale for selecting initial projects carefully.
2. List and explain the criteria for selecting initial projects.

3. What are the various components of a charter argument?

4. Explain the following concept: "Slow and steady is best."

ENDNOTES

1. R. Reedy, "Power Line Leukemia: Hot Issue of the '90s," *Safety and Health*, Vol. 139, No. 5 (May 1989), p. 59.

2. O. F. McDonald, "Noise: How Much Is Too Much?" *Safety and Health*, Vol. 144, No. 5 (November 1987), p. 37.

Establish, Train, and Activate Improvement Project Teams

- Team Composition
- Sample Improvement Project Teams (IPTs)
- Relationship of IPTs to the Steering Committee
- Criteria for Team Memberships
- Training Curriculum for Teams
- Timing of Training
- Team Activation
- Team Facilitation
- Step 13 in Action

The twelve steps leading up to this point involved planning, preparation, identification, and assessment. In the broadest sense, all twelve of the previous steps were preparation for this step. Step 13 is the most functional of the fifteen steps in the TSM implementation process. It involves establishing, training, and activating IPTs which, along with the Steering Committee and Facilitator, make up the *TSM Triad* (Figure 13–1), or the heart, brain, and soul of TSM.

TEAM COMPOSITION

IPTs are either *natural work teams* or *cross-functional teams*. A natural work team is composed of members who work closely together in the performance of their regular jobs. For example, in a manufacturing organization the CAD/CAM unit and the precision machining unit would both be natural work teams. An IPT that draws its members from either the CAD/CAM unit or the precision machining unit would be a natural work team.

Figure 13–1
Components of the TSM Triad

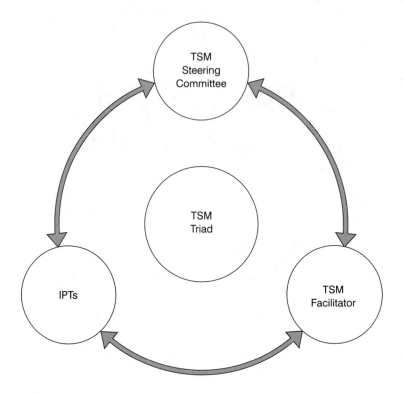

Cross-functional teams draw their members from several different departments or units. The members do not work closely together on a day-to-day basis in their regular jobs, but they all share a common interest in seeing a specific hazardous situation eliminated. Consequently, although members of cross-functional teams come from different organizational disciplines, they are all stakeholders (individuals who have a stake in a given situation).

This concept of the stakeholder is important when appointing employees to membership on an IPT. Regardless of whether the team is natural or cross-functional, all of its members should be stakeholders. Whether an IPT should be natural or cross-functional depends on the range of the hazardous situation in question. If the potential harmful effects of the hazard are limited to just one unit, a natural work team is typically the most appropriate configuration.

If the potential harmful effects of the hazard are likely to spread beyond a given unit, a cross-functional team representing all stakeholders is the recommended approach. Since it is the nature of hazards to cross departmental boundaries, more often than not IPTs are cross-functional in their composition.

In the previous step, the need to choose carefully when selecting initial projects was stressed. In this step it is important to apply an equal measure of care in selecting team members. Regardless of whether the team is natural or cross-functional, all team members should be drawn from the list of advocates that is part of the tailored plan developed in Step 11. This requirement makes it difficult to use the natural-work-team approach when activating initial teams because there may be an insufficient number of advocates in a given unit during the early stages of implementation. As projects are undertaken and successfully completed, the number of advocates will increase to the point where after as little as a year most organizations will have both natural and cross-functional teams activated and working simultaneously.

SAMPLE IMPROVEMENT PROJECT TEAMS (IPTS)

Over time most organizations find that both natural and cross-functional teams are needed. Figure 13–2 is an example of a natural work team established to pursue the problem of back injuries in the Shipping Department of an electronics manufacturing firm. Figure 13–3 is an example of a cross-functional team established to pursue the problem of carpal tunnel injuries among computer users.

The IPT depicted in Figure 13–2 consists of seven employees selected from the Shipping Department. The team captain was chosen, in this case, by the team members. Unless the Steering Committee has a specific individual in mind, it is best to let IPTs choose their own team captains. The other team members are loaders, packagers, or conveyor operators. Their jobs involve packaging, handling, and loading finished prod-

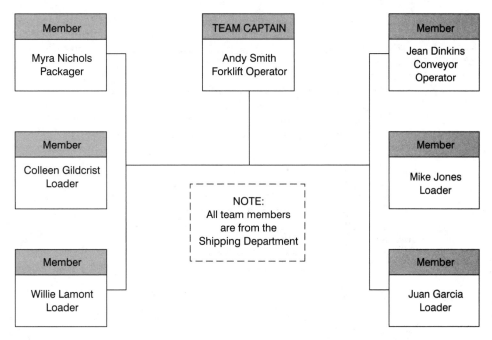

Figure 13–2
Natural Work Team (IPT)

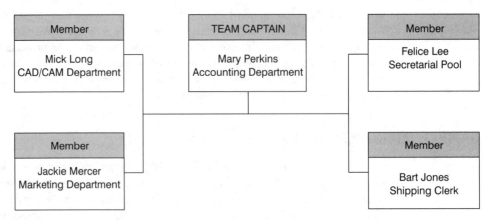

Figure 13–3
Cross-Functional Team

ucts for shipping to customers. They all know each other and work together regularly. Each member of this team is considered an advocate by the Steering Committee.

The IPT depicted in Figure 13–3 consists of five employees selected from five different departments in which there is heavy computer use. Consequently, this is a cross-functional team. The captain, Mary Perkins, was selected by her fellow team members. What these employees have in common is computer use. The computer is the primary tool of their jobs, and they use it constantly. In addition to being computer users, these five team members are considered advocates by the Steering Committee.

RELATIONSHIP OF IPTS TO THE STEERING COMMITTEE

IPTs, represented by their team captains, work directly with the TSM Facilitator on a day-to-day basis. However, ultimate responsibility for IPTs rests with the Steering Committee. Figure 13–4 depicts the relationships of IPTs to the TSM Facilitator, the Steering Committee, and each other. The Steering Committee is responsible for the following activities as they relate to IPTs (Figure 13–5):

- *Selecting projects* to be pursued by IPTs
- *Approving team charters* drafted by the TSM Facilitator
- *Selecting team members* for both natural work teams and cross-functional teams
- *Providing the resources* needed by IPTs in order to carry out their charters
- *Monitoring the progress* of IPTs in fulfilling their charters
- *Disbanding teams* when their work has been accomplished

The TSM Facilitator serves as an intermediary between the Steering Committee and IPTs. This individual works directly with IPTs, and keeps the Steering Committee informed concerning progress and problems. The Facilitator is of critical importance to

Figure 13–4
Relationships of TSM Entities

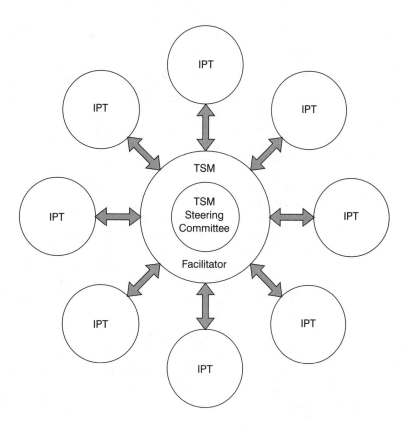

Figure 13–5
Functions of the Steering
Committee Pertaining to IPTs

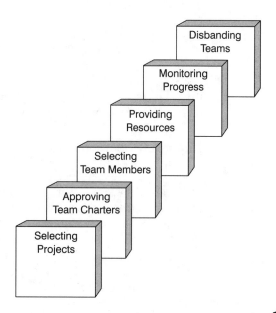

TSM TIP

Airborne Toxic Substances

Asphyxiants are substances that can disrupt breathing so severely as to cause suf-focation. Asphyxiants may be simple or chemical. Common simple asphyxiants include carbon dioxide, ethane, helium, hydrogen, methane, and nitrogen. Common chemical asphyxiants include carbon monoxide, hydrogen cyanide, and hydrogen sulfide.

---

the success of IPTs, and to the success of TSM. However, responsibility for IPTs and for TSM ultimately rests with the Steering Committee.

Consequently, the relationship of the Steering Committee to an IPT is the same as that of a manager to an employee. While managers may delegate work to employees, and empower them to carry it out, they are still responsible for the work. Work can be delegated. Authority can be delegated. But responsibility cannot.

CRITERIA FOR TEAM MEMBERSHIP

The necessity for selecting advocates when establishing IPTs has been explained at length. There are other important criteria for selecting team members. These criteria can be divided into two broad categories as follows: *mandatory* and *important*.

Mandatory Criteria

In addition to being an advocate, ideally team members will also be nonvolatile and unbiased. Volatility and bias are counterproductive in a team setting. Teamwork is built on trust, and it assumes that team members will work cooperatively toward common goals. Individuals who pursue personal agendas that are tainted by bias will not be trusted by their teammates. Individuals who behave in a volatile manner will antagonize their teammates in addition to betraying or failing to earn their trust.

Important Characteristics

In addition to the mandatory criteria, there are also several other characteristics that can be important for team members. These additional characteristics include the following: good interpersonal skills, good communication skills, open-mindedness, and personal interest. Personal interest exists when the individual in question is a stakeholder.

Good interpersonal skills help team members get along. While it is not necessary that team members be friends, it is necessary that they get along even when they disagree, which is a normal occurrence in a team setting. Interpersonal skills are the oil in the cogs of human discourse. They allow people to disagree without being disagreeable.

In a team the work of one member affects the success of all members. Think of a football team. On every play every team member has an assignment that must be carried out properly in order for the play to work. When the quarterback sees something he doesn't like or when an unexpected opportunity presents itself, he changes the called play by audibilizing (calling a new and different play at the line just before the ball is snapped). The audibled play will work only if it is effectively communicated to all players. In other words, all players need to hear, understand, and act on the quarterback's messages.

To enhance communication, the quarterback is plugged into an elaborate system that includes coaches on the sidelines and in observation boxes near the top of the stadium. The purpose of the system is to communicate accurate, up-to-date information that will enhance team performance and to do so constantly and effectively. Effective communication is just as important to members of IPTs as it is to members of professional football teams. The better the communication skills of team members, the better communication will be on the team.

Open-mindedness is an important characteristic for members of IPTs. The purpose of an IPT is to examine a problem and make recommendations for eliminating it. When undertaking such an assignment, the ability to look at a situation from every possible point of view will typically lead to both a more accurate description of root causes and a more comprehensive list of potential remedies.

Closed-minded team members often take a *don't-bore-me-with-the-facts-my-mind-is-made-up* approach to their work. Such an approach will throw a wet blanket on the investigative zeal of team members and limit their effectiveness in finding innovative solutions.

It is important that all members of an IPT have a personal interest in finding a workable solution to the problem at hand. In other words, all team members should be stakeholders. Personal interest is a strong motivator. By selecting only stakeholders as IPT members, the organization ensures a high level of motivation. In other words, this approach guarantees that all team members really want to see the problem in question solved.

TRAINING CURRICULUM

Once an IPT has been established and chartered, it should be trained. To activate a team without first helping its members develop fundamental knowledge and skills concerning the task at hand is a mistake. The charter tells the team what it is expected to do, but it does not and cannot tell the team how to do it. Imagine giving a class of children a list of books to read without first teaching them how to read. Activating an untrained team would be a comparable mistake. Figure 13–6 contains the outline for a curriculum that represents the minimum training for an IPT.

The curriculum shown in this figure consists of six hours of training. Consequently, it can be accomplished comfortably in just one work day with plenty of time for breaks, lunch, and informal interaction. Any facility that will comfortably house the number of employees in question will suffice for the training. However, to guard against interruptions and work-related distractions, the training should be undertaken off-site if possible.

Materials Processing Company	
Topic	**Time**
TSM Overview	**90 Minutes**
• TSM Defined	
• TSM Steering Committee	
• TSM Facilitator	
• Improvement Project Teams	
Fundamentals of Teamwork	**90 Minutes**
• Rationale for Teamwork	
• Trust and Trustbuilding	
• Team Performance vs. Individual Performance	
• Accountability	
Communication	**90 Minutes**
• The Communication Process and Its Role	
• Inhibitors of Communication	
• Listening Skills	
• Conflict Presentation and Resolution Techniques	
Hazard Identification	**90 Minutes**
• Identifying Root Causes	
• Hazard Identification Checklists	
• Recommending Solutions	

TSM Overview

The workshop begins with an overview of TSM to ensure that all participants understand the concept and where they fit into it. The TSM Facilitator is the trainer for this segment. This individual provides a working definition for TSM, explains the Steering Committee concept including both membership and role, describes his or her job as the TSM Facilitator, and finally, explains the concept and role of the IPT.

This segment of the training establishes the foundation for all that follows. Consequently, it is important for the Facilitator to approach the topic slowly, deliberately, and with patience, allowing plenty of time for questions, discussion, and interaction among participants. Figures 13–7, 13–8, 13–9, and 13–10 are examples of materials that might

Figure 13–7
Sample Definition Sheet for
Training–TSM

Materials Processing Company

TSM Defined

Total Safety Management (TSM) is a performance-oriented
approach to safety and health management that gives our organization a sustainable competitive advantage in the global marketplace by establishing a safe and healthy work environment
that is conducive to consistent peak performance, and that is
improved continually forever.

Figure 13–8
Sample Definition Sheet—
Steering Committee

Materials Processing Company

TSM Steering Committee

MPC's TSM Steering Committee consists of the company's executive management team augmented by the Director of Safety and
Health. The Steering Committee is responsible for the formulation
of safety and health policies, approval of internal regulations and
work procedures relating to safety and health, allocation of
resources, and approval of recommendations by IPTs.

---

be used as handouts, overhead transparencies, slides, computer graphics, or some other
form of visual aid to instruction.

Fundamentals of Teamwork

The *fundamentals-of-teamwork* segment of the training is next. It should be conducted
by a teamwork expert, even if this requires bringing in an outside consultant. The first
topic in this segment is "Rationale for Teamwork." Figure 13–11 is an example of a
visual aid that might be used to explain the rationale for teamwork.

Notice that Figure 13–11 begins with a definition. There are three key elements in
this definition that should be understood by all members of IPTs. The first element is
"individuals who work together." On a team, the performance of the group is what matters, not the performance of individuals. Unilateral action is inappropriate. An individual's performance is judged on the basis of what he or she contributes to the performance of the group. This *team-first* mentality should be understood and accepted by all
team members from the outset.

The second element has to do with "mutually-supportive" interaction among team
members. In a team setting, individual members help each other. Think of a basketball

Figure 13–9
Sample Definition Sheet—
TSM Facilitator

Materials Processing Company
TSM Facilitator
MPC's Director of Safety and Health is the TSM Facilitator. This individual serves as the Steering Committee's resident expert on the technical and compliance aspects of safety and health, and is responsible for the overall implementation and operation of MPC's TSM program.

Figure 13–10
Sample Definition Sheet—IPTs

Materials Processing Company
Improvement Project Teams (IPTs)
IPTs are ad hoc or temporary teams of employees chartered by the TSM Steering Committee to pursue specific improvement projects relating to the work environment.

team whose individual players work well together. They help each other by setting up plays, passing, rebounding, setting picks, and making assists. Now think of a basketball team whose members take a self-serving *me-first* attitude. They fight each other for the ball as hard as they fight the opponent, and typically with predictable results. It is mutual support and working together that make a team greater than the sum of the individual talents of its members.

The third element is "collective goals." This element will do more than anything else to promote working together in a mutually supportive manner. It is a fundamental characteristic of human nature that people are drawn together by common goals. A common goal can bring together people who agree on little or nothing else but their common goal. Common goals can transcend gender- or age-based, racial, cultural, philosophical, religious, and political differences.

Consequently, it is important for members to understand the team's goals and accept them. The charter ensures that all members understand the team's goals. Naming only stakeholders as team members helps ensure acceptance of them.

Explaining the various elements of the rationale for teamwork shown in Figure 13–11 helps members understand why safety and health problems are pursued by teams. Clearly, the organization wants the benefit of different opinions, perspectives, points-of-view, talents, and experience bases before implementing solutions to safety and health problems. Just as a basketball team whose members work well together is more likely to perform better than five self-serving, uncoordinated individuals, an IPT whose members work well together is more likely to perform better than individual employees.

Figure 13–11
Sample Visual Aid for TSM
Training—Teamwork Rationale

Materials Processing Company
What Is a Team? A team is a group of individuals who work together in a mutually supportive way to accomplish collective goals. **Rationale for Teamwork** • Two or more heads are better than one. • Two or more sets of eyes will see more than one. • Two or more sets of ears will hear more than one. • Teamwork promotes communication. • Because of synergism, a team is greater than the sum of its individual members.

Communication

The next component in the curriculum deals with communication. This segment should be conducted by a communications expert even if it is necessary to bring in an outside consultant. All IPT members need to understand that communication is a process, that telling is not communicating and hearing is not understanding. Figure 13–12 is an example of a visual aid that might be used during this segment of the training.

Figure 13–12
Sample Visual Aid for TSM
Training—Essentials of
Communication

Materials Processing Company
Communication Defined Communication is the transfer of a message (information, idea, emotion, intent, or feeling) that is both received and *understood*. **Communication Process** Sender ⟷ Message ⟷ Receiver **Effective Communication Defined** Effective communication is the transfer of a message received, understood, and *acted on in the desired manner*.

This visual shows that communication is a process that consists of four distinct components: (a) a sender, (b) the message that is sent, (c) the medium by which the message is sent (e.g., voice, telephone, E-mail, writing, facsimile), and (d) a receiver. In order for communication to occur, the message that is sent must be accurately received and understood. Have you ever been misunderstood? Have you ever misunderstood someone? For most people the answer to both of these questions is the same: "Yes, many times." In fact, every time people attempt to communicate there are a variety of factors working against them.

Factors associated with the sender include such shortcomings as poor speaking skills, confusing nonverbal cues, and poor writing skills. Factors associated with the medium include such problems as noise, computer glitches, electronic interference, and equipment malfunctions. Factors associated with the receiver include poor listening skills, insensitivity to nonverbal cues, underdeveloped reading skills, and inability to properly use technological devices (computer, facsimile machine, video cassette player, audio cassette player). The inhibitors of communication mentioned herein as well as several others are summarized in Figure 13–13. This figure is an example of another visual that might be used during this segment of the training.

Listening is critical since most communication involves the spoken message delivered either in person or via some technological medium. Figures 13–14, 13–15, and 13–16 are visuals that might be used in conjunction with instruction on listening. Members of IPTs need to understand the various barriers to effective learning summarized in Figure 13–14. Being aware of these barriers and taking the steps necessary to overcome them will help members improve the effectiveness of their learning.

Figure 13–13
Training Aid—Inhibitors of Communication

Materials Processing Company

Inhibitors of Communication

- Difference in meaning
- Lack of trust
- Information overload
- Interference
- Condescending tone
- Poor receiving skills (listening, reading, technological)
- Premature judgments
- Inaccurate assumptions
- Kill-the-messenger syndrome
- Cultural complications

Figure 13–14
Sample Checklist—Effective
Listening Barriers

Materials Processing Company
Barriers to Effective Listening
✓ Lack of concentration
✓ Interruptions
✓ Distractions
✓ Preconceived notions
✓ Tuning out
✓ Thinking ahead
✓ Interference

Figure 13–15
Sample Checklist—Improving
Listening

Materials Processing Company
Listening Improvement Checklist
✓ Remove distractions
✓ Put the speaker at ease
✓ Look directly at the speaker
✓ Concentrate
✓ Observe nonverbal cues
✓ Be patient and wait
✓ Don't interrupt
✓ Ask questions for clarification
✓ Paraphrase and repeat
✓ Control your emotions no matter what is said

Figure 13–15 contains specific strategies that can be used to improve listening. These are strategies over which listeners have control and which they can apply at will. Members who learn to apply these strategies will be more effective listeners. Figure 13–16 contains the characteristics of responsive listeners. Responsive listeners listen not just with their ears, but also with their eyes, brains, and hearts. Responsive listening is an *active* undertaking in which an individual seeks to accurately receive the message, understand it, and affirm that understanding. Responsive listeners are typically the most effective listeners.

Figure 13–16
Sample Checklist—Responsive
Listening

> **Materials Processing Company**
>
> **Responsive Listening Characteristics**
> Responsive listeners have the following
> characteristics:
>
> ✓ Active
> ✓ Alert
> ✓ Vigilant
> ✓ Sensitive

Conflict prevention and resolution are important skills for members of an IPT or any other type of team. People who work in teams will occasionally disagree. This is normal, natural, and to be expected. Disagreements among team members become a problem only when those who disagree become disagreeable. This is when conflict occurs, and conflict, if left unresolved, can undermine the efforts of a team. Figure 13–17 is an example of a visual that might be used when teaching conflict prevention and resolution skills.

All members of IPTs need to understand the various causes of conflict in the workplace as a first step to knowing how to prevent and/or resolve conflict. Conflict over limited resources is likely to occur when an IPT recommends a solution that is perceived as costing one department an amount that is out of proportion compared with costs to other departments. The charter should solve any problems that otherwise might arise relating to incompatible goals. This is another reason why it is so important to have a charter, and for all members of the IPT to understand and accept the goals set forth in it.

Role ambiguity within a team is one of the most common causes of conflict. Imagine a football team in which the players are not sure of who is supposed to play what

Figure 13–17
Visual Aid—Causes of Conflict

> **Materials Processing Company**
>
> **Causes of Workplace Conflict**
> • Limited resources
> • Incompatible goals
> • Role ambiguity
> • Different values
> • Different perspectives
> • Communication problems

position. "I-am-the-quarterback" arguments would occur on every play, and before long, there would be chaos. The same can be said for IPTs. This is why it is important for an IPT to select a team captain and for that person to clearly define the roles of other team members.

The last three factors in Figure 13–17 are the most common causes of conflict in the workplace. People with different values see things differently. This is illustrated every day when people at different places on the political spectrum from liberal to conservative look at the same issue, but see it differently. Even having common goals cannot always overcome the challenge of different values. Different values tend to give people different perspectives. For this reason it is important for members of teams to learn to be patient, open-minded, and sensitive to views that differ from theirs. Communication problems just exacerbate conflict that grows out of differences in values and perspectives. On the other hand, good communication skills can overcome conflict that is rooted in these factors.

Figure 13–18 contains a checklist of strategies team members can use to resolve conflict when it occurs. Using these strategies, team members can resolve conflicts that occur among them without involving the team captain, supervisors, or higher level managers. This is the preferred approach. Any time an external agent must be called on to help resolve conflict between or among team members, a certain amount of trust is lost, and teamwork is built on trust.

Figure 13–18
Strategies for Defusing Conflicts

Materials Processing Company

- If you feel yourself or the other person involved becoming angry, suggest a "time-out" to cool down before proceeding.
- Determine how important the issue in question is (to you and to the other party involved).
- If the issue is important enough to warrant further discussion, find a quiet, private place to talk about it.
- Agree to some groundrules such as the following: (a) This is important to both of us so let's agree to approach the issue in an open, honest, positive manner; (b) Let's agree that *we* are responsible for finding an agreeable solution; and (c) Let's agree to be mutually supportive and nonaccusatory in looking for solutions.
- Work together to clearly define the real problem.
- Let the other individual propose solutions, and you do the same. Don't fall into the "If-it's-not-my-idea-it's-a-bad-idea" trap.
- Subject all proposed solutions to an informal cost/benefit analysis. Select the solution or combination of solutions that makes the most sense.

Hazard Identification

The final segment of an IPT's basic training covers the topic of hazard identification. It involves helping members learn how to identify hazards in the work environment, determine the root cause of the hazard, and make recommendations for eliminating the cause.

It should be noted here that we are not talking about the type of detailed hazard analysis that involves the use of such methods as failure mode and effects of analysis (FMEA), hazard and operability review (HAZOP), technique of operations (TOR), human error analysis (HEA), fault tree analysis (FTA), or risk analysis. These methods are reserved for the use of safety and health professionals, and cannot be adequately taught to nonprofessionals in a short workshop.

IPT members learn to use hazard identification checklists prepared for them by safety and health professionals (Figure 13–19) and a root cause identification tool such as a Fishbone diagram (Figure 13–20).

Materials Processing Company
Hazard Identification Checklist
Noise/Vibration

- What are the sound level meter readings taken at different times and locations?
- What are the dosimeter readings for each shift of work?
- Are audiometer tests being conducted at appropriate intervals?
- What are the results of audiometric tests?
- What is being done to reduce noise at the source?
- What is being done to reduce noise along its path?
- Are workers wearing personal protective devices where appropriate?
- Are low-vibration tools being used wherever possible?
- What administrative controls are being used to limit employee exposure to vibration?
- Are employees who use vibrating tools or equipment doing the following as appropriate:
  - Wearing properly fitting thick gloves?
  - Taking periodic breaks?
  - Using a loose grip on vibrating tools?
  - Using vibration-absorbing floor mats and seat covers as appropriate?

Figure 13–19
Sample Hazard Identification Checklist

Figure 13–20
Fishbone Diagram Using the
Five M's

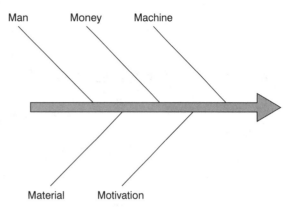

Figure 13–19 is a checklist prepared by a safety and health professional to help an IPT identify noise and vibration hazards in the workplace. In proceeding through the checklist, IPT members are free to request as much assistance from the TSM Facilitator as they might need in order to satisfy their team charter. For example, assistance might be needed in reviewing the results of audiometric tests or in taking dosimeter readings. Similar checklists for other areas of concern (automation, ergonomics, electrical, toxic substances, etc.) are contained in the Appendix.

Once an IPT has identified a potential problem, the next step is to identify the root causes of the problem. For example, if by working through the checklist the IPT determines that employees are not wearing appropriate personal protection devices in high-noise areas, its members will want to determine why. Is it because they don't have access to personal protective devices, don't know how to use them, or just don't want to use them?

A tool such as the one shown in Figure 13–20 can help narrow such questions down until the root cause is determined. The Fishbone diagram in this figure is already set up with the *Five M's* (Man, Money, Machine, Material, and Motivation). These five areas, although certainly not all-inclusive, do typically represent a good starting point when looking for the causes of workplace problems. Additional spines may be added as necessary.

For example, when dealing with safety and health problems, it is usually good to add a spine for management. This spine is used for recording administrative controls that are missing or that are present, but are being overlooked, neglected, and/or poorly enforced.

As IPT members interview employees, they record what is learned on Fishbone diagrams along the appropriate spines. Figure 13–21 shows what the final Fishbone might look like after the IPT has looked into the question of why employees are not wearing personal protective devices in high-noise areas. This is a composite diagram that reflects what was learned by all members of the IPT.

In the category of *man* (staff of whatever sex, of course), two contributing factors were identified. First, many employees don't wear their eye-protection devices because

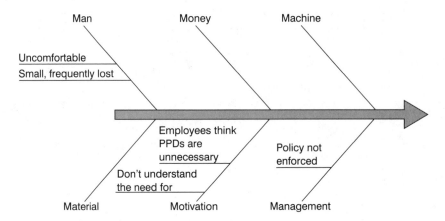

Figure 13–21
Sample Final Fishbone Diagram

they are uncomfortable. Second, devices are issued to individual employees. Because the devices are small they can be easily misplaced, and frequently are. The fact that the devices are uncomfortable probably contributes to the frequency with which they are lost.

In the category of *motivation*, team members found that their fellow employees view ear-protection devices as being unnecessary; they don't understand the need for them. Clearly, employees are not well informed concerning the potential for hearing loss from frequent and prolonged exposure to noise hazards.

In the category of *management*, team members found that their fellow employees just disregard the organization's ear-protection policy because they know it goes unenforced. The prevailing opinion among employees seemed to be "Why should we take ear protection seriously when management clearly doesn't?" There were no contributing factors found in the categories of *money, material,* or *machine*.

With all contributing causal factors identified, the IPT would forward a set of recommendations to the TSM Facilitator. Recommendations relating to the causal factors in Figure 13–21 might include the following:

- Identify more comfortable ear protection devices. Let employees field-test them.
- Make ear-protection devices readily available at numerous locations in the plant instead of issuing them to employees. Perhaps a disposable device can be obtained.
- Provide employees who work in high noise areas with training/information they need in order to understand why ear protection is important.
- Assign enforcement of the ear-protection policy to the supervisors of employees who work in high-noise areas. Have management conduct periodic spot checks to ensure supervisors are enforcing the policy and employees are following it.

TSM TIP

Electrical Hazards

A major cause of electrical shock in the workplace is contact with a bare wire that carries current. Employees should be taught to look for wire around which the insulation has deteriorated, leaving bare wire exposed.

TIMING OF TRAINING

A frequently asked question concerning the implementation of TSM is "Why not train all employees at the same time?" Proponents of the everybody-at-once approach typically justify their advocacy by stating the following reasons for it:

- Time savings
- Economy of scale
- Ensures that all employees are ready even before being assigned to an IPT

Saving time and economy of scale have a certain appeal when viewed strictly from a logistical and administrative perspective. The problem with these reasons is that they confuse efficiency and effectiveness. Saving time and gaining economy of scale might make the training process more *efficient*, but they probably won't make it more effective.

The most effective way the author has found to provide training is on a just-in-time basis or, in other words, just before it is actually needed. This may not be as efficient from a logistical point of view, but it is more effective from a teaching-and-learning perspective. Just-in-time training offers several advantages that make it more effective than the all-at-once approach. These advantages are as follows:

- Just-in-time training is fresh when it is applied in a live setting. Employees trained all at once may forget what they learn before being assigned to an IPT.
- Just-in-time training can be tailored to the specific team or teams in question. All-at-once training must necessarily be generic in nature.
- Just-in-time training has the benefit of *application proximity*. Because the need to apply what is being learned is just around the corner so to speak, participants tend to take the training more seriously than their all-at-once counterparts who don't know when they might get a chance to apply what they are learning.

Application proximity negates the third reason given by all-at-once advocates. When employees are trained all-at-once there is typically a span of time between the end of training and the beginning of live application. The longer this span of time is, the more likely it is that employees will have forgotten what was learned.

A good rule of thumb is "Train as many employees on a just-in-time basis as can be immediately assigned to teams that are at the point of being activated." This approach will ensure that the maximum benefit is gained from the training.

TEAM ACTIVATION

At this point in the process an IPT has been chartered, its members have been selected, and the new team has been trained. All that remains is to activate the team. Team activation is a substep in the overall implementation process undertaken to ensure that team members understand what they are supposed to do, as well as the boundaries within which the team must work, reporting procedures, and projected time frame milestones.

These issues are dealt with in the team's charter. But even with a written charter, it is still a good idea to have a face-to-face kick-off session. Such an orientation gives the TSM Facilitator an opportunity to double check the understanding of team members, and clear up any questions that remain.

The team activation meeting is conducted by the TSM Facilitator. Figure 13–22 summarizes the topics that should be reviewed during the team activation meeting. The

Figure 13–22
Topics to Review upon
Activation of IPT

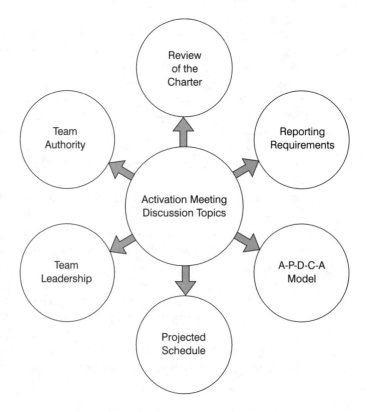

most important aspect of the meeting consists of the questions members ask. The review of the charter focuses on the goals contained in it. Reporting requirements are discussed next. Who does the reporting, in what form, to whom, and how often? These issues are discussed and clarified to the satisfaction of all involved. Then the A-P-D-C-A Model is reviewed one more time to ensure that all team members know how they will apply it. The team is chartered to conduct the *Assessment* step and make recommendations that are the basics for the *Planning* step. The *Do*, *Check*, and *Adjust* steps occur as the Steering Committee implements the team's recommendations.

Tentative deadlines for individual milestones and the overall time frame for the project are reviewed next. At this point in the discussion, the TSM Facilitator should explain what the team is to do when deadlines cannot be met. The team captain's function is reviewed next, with special emphasis on the relationship of the captain and the Facilitator, the captain and individual team members, and the team captain and the overall team.

The activation meeting concludes with a review of the team's authority. Typically, the team's authority is limited to recommending solutions. Implementing solutions is the province of the Steering Committee. If the team discovers a hazard that has the potential for immediate danger or damage, the TSM Facilitator is notified and corrective measures are implemented right away.

===== TSM CASE STUDY =====

What Would You Do?

"Why do we need an activation meeting? We've chartered the IPT. Let it get to work." John Moses, CEO of the Briarwood Company cannot understand why his TSM Facilitator, Monica Gates, wants to have a team activation meeting. As far as he is concerned, the time has come to see some action. Gates isn't sure what to tell her boss. If you were in her place, what would you tell Moses?

TSM FACILITATION

The TSM Facilitator is the team's link to the Steering Committee, as shown in Figure 13–23. The Facilitator monitors the activities of all active IPTs and reports to the Steering Committee on a regular basis concerning progress, problems, and recommendations. Correspondingly, the captains of all active IPTs make either verbal or written reports to the TSM Facilitator on a regular basis. As shown by Figure 13–23, communication in all levels of the model is two-way in nature.

When the Facilitator reports to the Steering Committee concerning the progress, problems, or recommendations of an IPT, the Committee's response is promptly reported back to the IPT. When the Steering Committee has questions or information for an IPT, the TSM Facilitator talks immediately with the captain of the team in question, and reports promptly back to the Steering Committee.

Figure 13–23
Reporting under TSM

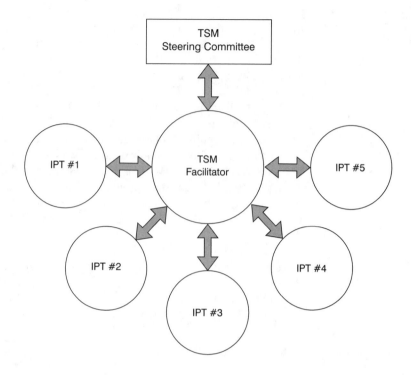

The Facilitator's role with regard to teams involves the following types of activities (Figure 13–24):

- Serving as the *communication link* between the Steering Committee and IPTs, and among IPTs.
- Serving as *advocate* when an IPT makes requests or recommendations to the Steering Committee.
- Serving as a *filter* to ensure that recommendations sent to the Steering Committee by IPTs don't violate OSHA regulations or any other applicable safety and health regulations.
- Serving as a *motivator* to encourage team members when their tasks become difficult.
- Serving as a *problem solver* when teams confront unexpected inhibitors and roadblocks.
- Serving as an in-house *safety and health consultant* when teams deal with technical issues or issues relating to regulatory compliance.

=================== STEP 13 IN ACTION ===================

Mack Parmentier is beginning to feel like a coach on the verge of a major victory. The game is not over, but everything is falling into place like it should. MPC's first IPT has

Figure 13–24
TSM Facilitator's Roles vis-a-vis
the IPTs

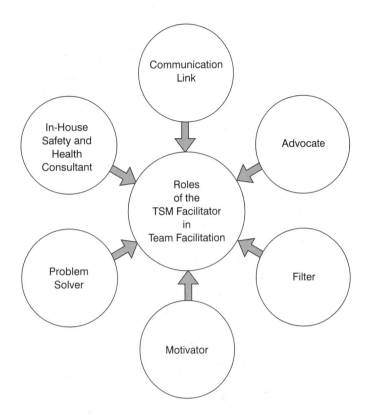

been selected and chartered. All members of the team are advocates of TSM, and the captain they selected is a solid performer who is widely respected throughout the company.

Team training had gone well. Parmentier had given the team an overview of TSM that included the following topics: TSM defined, role of the Steering Committee, role of the TSM Facilitator, and role of the IPT. Including questions, answers, and discussion, this segment of the training had taken just under 90 minutes to complete.

Parmentier had used an outside consultant to present the teamwork and communication segments of the IPT's basic training. Teamwork was dealt with before lunch, and communication right after. The consultant had done an excellent job. His simulation and role-playing activities had been well received and effective. Parmentier had completed the team's training with a 90 minute segment on hazard identification. The training focused on how to use a series of hazard identification checklists to root out safety and health problems in the workplace. Now all that remained was to activate MPC's first IPT and the company would be using TSM.

Parmentier intended to conduct the activation meeting himself the next day. During this meeting, he would review the team's charter and reporting requirements to make sure all members know what is expected of them. The A-P-D-C-A model will be reviewed one more time along with the projected schedule. Since the IPT has already selected its captain, Parmentier will touch only lightly on team leadership. His main focus will be to make sure that all team members understand the role of the captain and their relationship to him. Finally, Parmentier will review the issue of team authority with special emphasis on how to make recommendations.

As soon as the activation meeting is over, Parmentier will tell the team captain to get his team to work. From that point on, Parmentier's role will become one of team facilitation. He can't wait!

SUMMARY

1. IPTs may be either natural work teams or cross-functional teams. A natural work team is composed of members who work together every day in the performance of their jobs. Cross-functional teams consist of members from several different departments. A natural work team is used when the potential harmful effects of a hazard are limited to just the department or unit. A cross-functional team is used when the potential harmful effects of a hazard cross departmental boundaries.

2. Ultimate responsibility for the work of IPTs rests with the Steering Committee. The Steering Committee is responsible for selecting projects, approving team charters, selecting team members, providing resources, monitoring team progress, and disbanding teams that have fulfilled their charters.

3. Criteria for membership on an IPT fall into two categories; mandatory and important. Mandatory criteria include the following: TSM advocate, nonvolatile personality, and unbiased. Important criteria include the following: good interpersonal skills, good communication skills, open-minded, and interested.

4. Basic training for an IPT consists of four broad segments: TSM overview, fundamentals of teamwork, communication, and hazard identification. Training is most effective when provided on a just-in-time basis.

5. An IPT activation meeting should include a discussion of at least the following topics: review of the charter, reporting requirements, A-P-D-C-A model, projected schedule, team leadership, and team authority.

6. The TSM Facilitator helps IPTs succeed by serving as a communication link, advocate, filter, motivator, problem solver, and consultant.

KEY TERMS AND CONCEPTS

Advocate	Just-in-time training
Communication	Natural work team
Fundamentals of teamwork	Nonvolatile
Good communication skills	Open-mindedness
Good interpersonal skills	Personal interest
Hazard identification	Team activation

Team facilitation   TSM Triad
TSM overview   Unbiased

REVIEW QUESTIONS

1. Define the term *natural work team* and explain when one would be chosen for an IPT.
2. Define the term *cross-functional team* and explain when one would be used as an IPT.
3. Explain the relationship of an IPT to the Steering Committee.
4. List and explain the mandatory and important criteria for IPT membership.
5. Outline a basic training curriculum for an IPT.
6. Explain the concept of just-in-time training and why it is important.
7. List and explain the topics that should be discussed during an IPT activation meeting.
8. Explain the various roles the TSM Facilitator plays in facilitating the work of an IPT.

Activate the Feedback Loop

With Step 13 concluded, the organization now has IPTs at work throughout the organization. At this point, feedback becomes critical. The Steering Committee needs to know what the IPTs are doing, the extent of their progress, and the problems they are encountering. IPTs need to know what the Steering Committee thinks of its recommendations. The Steering Committee needs to know what employees think about improvements that have been made. This step focuses exclusively on how to use the feedback loop to maximize communication among the Steering Committee, IPTs, and employees.

FEEDBACK LOOP DEFINED

The Assess-Plan-Do-Check-Adjust model is a fundamental component of TSM. The Steering Committee applies the model at its level, and the IPTs apply it at theirs. Feedback occurs between the two groups as they each apply the model at their respective levels, as pictured in Figure 14–1.

The feedback loop consists of the various communication vehicles that connect the Check component of the A-P-D-C-A model with the Plan component. The Check compo-

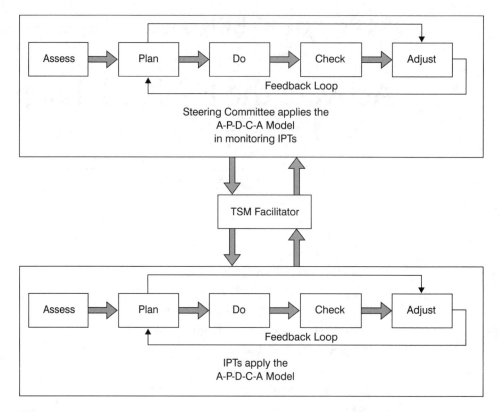

Figure 14–1
Two Applications of the A-P-D-C-A Model

nent answers the question "Is the plan we implemented working?" If the answer to this question is anything other than "Yes," the plan is adjusted as needed. The adjusted plan is then implemented, and the new results are checked. This process occurs as a continuous loop until the problem in question is corrected.

IPT FEEDBACK TO THE STEERING COMMITTEE

The TSM Facilitator maintains constant communication with all operational IPTs and passes on what he or she learns to the Steering Committee. There are several reasons why feedback from IPTs to the Steering Committee is important. These reasons include the following:

■   *The Steering Committee is ultimately responsible for safety and health in the organization.* The day-to-day work associated with safety and health is delegated to the organization's chief health and safety professional, who, typically, serves as the TSM

TSM TIP

Aids Education Programs

A well-designed AIDS education program can serve several positive purposes, including the following: (1) give management the facts needed to develop policy and make informed decisions; (2) change employee behavior in ways that make them less likely to contract or spread the disease; (3) prepare an organization to respond properly when an employee contracts the disease; and (4) decrease the likelihood of legal problems resulting from inappropriate response to an AIDS-related issue.[1]

Facilitator. This individual receives assistance from IPTs and any other safety and health professionals who may be on staff. Safety and health professionals and IPTs perform their duties on behalf of the Steering Committee, which retains ultimate responsibility for safety, health, and all other related organizational concerns. The Steering Committee can delegate the authority necessary for people to do specific jobs, but not the responsibility for ensuring that those jobs are properly carried out. For this reason, it is essential that the Steering Committee receive accurate and continual feedback.

■ *New teams are unknown, inexperienced entities.* When TSM is still young in an organization, IPTs will be inexperienced and untried. Are the individual members able to work together as a team? Do they need help the Steering Committee can provide? These and other questions that come up are answered by feedback. Inexperienced teams are sure to encounter problems because even experienced teams do. The Steering Committee can probably solve or help solve any problems an IPT will confront, but only if it knows about them.

■ *The Steering Committee must ensure that the charter is being carried out.* IPTs receive a charter from the Steering Committee that is both definitive and specific. Certain actions are supposed to be accomplished by specified dates, or, at least, within a well-defined time frame. The Steering Committee needs to know that the charter is being carried out as intended or if not, why not. This requires accurate feedback provided on a regular basis.

FREQUENCY AND FORMAT OF FEEDBACK

The frequency of feedback from IPTs to the Steering Committee is a function of two factors: (a) experience of the IPT members, and (b) gravity of the problem the IPT is dealing with. During the early development of TSM, when teams are still young and inexperienced, the frequency of feedback is more critical than it will be when TSM has matured in the organization.

Consequently, during the formative stages frequent feedback is especially important. Frequent at this point should be viewed as meaning at least weekly. This means that at least once each week the IPT captain meets with the TSM Facilitator who, in turn, meets with the Steering Committee. Feedback provided in both cases centers on the team's charter, with special attention given to required tasks and their corresponding deadlines. Organizations in which TSM is mature and teams are experienced might require only monthly reports. These guidelines apply when IPTs are dealing with routine problems. However, when a team is dealing with an especially critical problem—for example, when the potential for injury or damage to property is imminent—the frequency of reporting might be daily regardless of the team's perceived level of competence.

Verbal reports are recommended as opposed to written. TSM is a real-time do-it-now type of concept. Written reports take time to prepare and can be left to languish in unattended in-boxes. This is not to say that verbal reports should not be supported by written and/or graphic materials. In fact, they should be. The point is that such materials should be used in addition to, rather than instead of verbal reports.

FEEDBACK FROM THE STEERING COMMITTEE TO IPTS

Feedback must occur in both directions: from IPTs to the Steering Committee, and vice-versa. Prompt feedback from the Steering Committee to IPTs is critical. When requests, questions, or recommendations are directed to the Steering Committee, the IPT should receive a prompt response. There are several reasons for this.

Often an IPT will be unable to proceed beyond a certain point until it receives feedback from the Steering Committee. For example, an IPT might recommend a next step that requires permission or the commitment of resources. In such a case, the team is on hold until it hears back from the Steering Committee.

Another reason why prompt feedback from the Steering Committee is important is that team morale can suffer for want of feedback. Employees know that managers deal with the issues they consider most important. Consequently, failure on the part of the Steering Committee to give an IPT prompt feedback can be the same as saying, "Your recommendation is not important enough to warrant our attention." IPT members who receive such a message will be reluctant to put any more effort into the team's assignment.

A final reason for prompt feedback to IPTs has to do with timeliness of improvements. Remember, IPTs are chartered to make the workplace safer and, in turn, more productive. If an IPT's recommendations are dealt with in a timely manner, improvements can also be made in a timely manner. The organization needs this to happen. It is why the TSM concept was adopted in the first place.

It should be noted that prompt feedback—which the author defines as within 48 hours—does not necessarily mean approval. Feedback can take any one of a number of forms including the following (see Figure 14–2):

■ *Acknowledgment.* This type of feedback tells the team, "We received your recommendation (question, request, etc.) and will have an answer for you no later than . . . " Such a response serves an important purpose. It lets the team know that its

Figure 14–2
Types of Feedback to an IPT

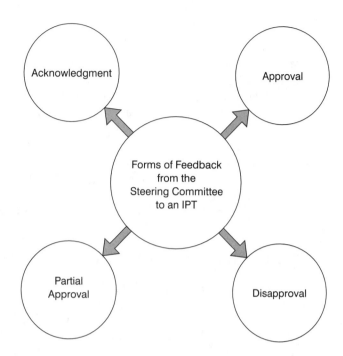

recommendation has, in fact, been received and will be acted on within a specified period of time. This closes the loop so that team members are not left wondering "Did the Steering Committee receive our recommendation or did it get lost in someone's in-basket? Does silence mean the Steering Committee is ignoring our recommendation?" If left to wonder about the status of a recommendation, people tend to think the worst.

TSM TIP

Cleaning Up Air and Water Pollution[2]

Since 1970, significant progress has been made in the United States in reducing air and water pollution. Consider the following evidence:

- *Carbon monoxide emissions have been reduced by more than 38 percent.*
- *Sulphur oxides are down by more than 25 percent.*
- *Ocean dumping of industrial wastes is down by 94 percent.*
- *Cities without adequate sewage treatment are down by 80 percent.*
- *Miles of river unfit for swimming are down by 44 percent.*

- *Approval.* This type of feedback is the simplest, and, from the IPTs's perspective, the most welcome. It says, "Your recommendation is approved. Proceed."
- *Partial approval.* It is sometimes necessary to approve only part of an IPT's recommendation. When this is the case, the Steering Committee should do the following: (a) be specific as to which aspects of the recommendation are approved and which are not; (b) explain why certain aspects of the recommendation were approved; (c) explain why other aspects of the recommendation were disapproved; and (d) make suggestions concerning those aspects of the recommendation that were disapproved.
- *Disapproval.* It will be necessary at times to reject an IPT's recommendation. When this is the case, the Steering Committee should proceed as follows: (a) inform the IPT promptly of the disapproval; (b) explain why the recommendation has been rejected; and (c) counsel the IPT concerning more feasible options it might consider.

EMPLOYEE FEEDBACK CONCERNING IMPROVEMENTS

Employee feedback is important in a TSM setting. An initial survey of employee attitudes concerning the state of the work environment was conducted in Step 10. The findings of this survey should be used as a baseline against which improvements are measured. Now that IPTs, the TSM Facilitator, and the Steering Committee are busy working together to improve the work environment, employee attitudes should improve correspondingly.

But it would be a mistake to just assume that employee attitudes are improving. In a TSM setting, a good rule of thumb to follow is: "Don't assume, measure." What do employees think of the improvements? Have the improvements had a positive impact on their performance? Were the improvements made the ones that were really needed? The Steering Committee needs to have answers to these questions. Consequently, employee feedback is critical.

VALUE OF EMPLOYEE FEEDBACK

The link between employee perceptions of the work environment and employee performance was spoken to in the narrative for Step 10. Employees who are unsettled by perceived threats from safety and health hazards in their work environment are not likely to perform consistently at peak levels. Peak performance in any job requires total concentration. Mental energy diverted by workplace hazards is energy that could have been focused on improving performance.

Employee feedback collected periodically on a formal basis will tell the Steering Committee something it has got to know: whether employees are focusing on improving their performance or on the workplace hazards they perceive as threats. If employee perceptions have improved, the Steering Committee will know that its efforts are working. If improvements have been made, but employee perceptions have not improved correspondingly, the Steering Committee also needs to know this. In such cases, the Steering Committee will want to ask itself the following questions:

- Do employees know that the improvements were made?
- Were the improvements made those that most directly affect employees on a day-to-day basis?
- Has sufficient time elapsed for employees to recognize a difference?

Employee feedback has value—particularly when coupled with hard data that is collected on a regular basis—in answering the following question: "Are our improvements making a difference?" In helping answer this question, employee feedback becomes an important part of the Check step in the Steering Committee's application of the A-P-D-C-A cycle.

FEEDBACK COLLECTION METHODS

In Step 10, several methods were introduced for collecting initial employee feedback. These methods were the survey (internal), focus group (internal), focus group (external) and survey (external). The same methods should be used for collecting subsequent feedback. They are reviewed briefly in the following subsections.

Survey (Internal)

The most widely used method for collecting employee feedback concerning the work environment is the employee survey conducted as an in-house project. With this method, a survey instrument such as the one shown in Figure 14–3 is developed by the TSM Facilitator. The actual instrument is tailored specifically for the types of improvements the organization has made.

The survey instrument is distributed to all employees who are stakeholders concerning the improvements in question along with a cover memorandum explaining its purpose, the time frame within which it should be completed, and how confidentiality will be maintained. Once all responses have been collected, they are summarized and the individual forms are destroyed.

Focus Group (Internal)

Another widely used method for assessing employee perceptions as they relate to the work environment is the focus group conducted as an in-house project. With this method, employees representing all departments and units that are stakeholders are invited to be members of a focus group. In larger organizations, more than one group may be required because the size of a group should be between 10 and 15. With fewer participants involved, the group may represent too narrow a perspective. With more, the group may be unwieldy. As with the survey method, a document such as the one in Figure 14–3 is developed by the TSM Facilitator. This document is used to prompt discussion while simultaneously focusing it on concerns about the work environment.

Employee Perceptions Survey

Perception Statement	Strongly Disagree	Disagree	Agree	Strongly Agree
1. The workplace is free of unreasonable stress.	_____	_____	_____	_____
2. The workplace is free of mechanical hazards.	_____	_____	_____	_____
3. The workplace is free of falling and impact hazards.	_____	_____	_____	_____
4. The workplace is free of lifting hazards.	_____	_____	_____	_____
5. The workplace is free of heat/temperature hazards.	_____	_____	_____	_____
6. The workplace is free of electrical hazards.	_____	_____	_____	_____
7. The workplace is free of fire hazards.	_____	_____	_____	_____
8. The workplace is free of toxic substance hazards.	_____	_____	_____	_____
9. The workplace is free of explosive hazards.	_____	_____	_____	_____
10. The workplace is free of radiation hazards.	_____	_____	_____	_____
11. The workplace is free of noise/vibration hazards.	_____	_____	_____	_____
12. The workplace is free of automation hazards.	_____	_____	_____	_____
13. The workplace is free of bloodborne pathogen hazards.	_____	_____	_____	_____
14. The workplace is free of ergonomic hazards.	_____	_____	_____	_____
15. The workplace is free of violence hazards.	_____	_____	_____	_____

Figure 14–3
Sample Employee Survey

Survey (External)

This approach is the same as the internal survey with one exception: the survey is conducted and summarized by an outside agent. Using an external consultant can relieve the anxiety of employees about confidentiality. It can also solve the time consumption problem experienced by large companies when summarizing survey input. The external survey, like the internal, involves all employees who are stakeholders.

Focus Group (External)

This approach is the same as the internal focus group with one exception: the focus group sessions are conducted by an external consultant. Using an external consultant can lessen any insecurities the employees may have about confidentiality issues. Employees who don't yet have a high level of trust toward management may be more willing to open up to an external consultant.

PUTTING EMPLOYEE FEEDBACK TO USE

Employee feedback has value for an organization if and only if it is put to proper use. Some of the ways it can be put to good use are as follows (see Figure 14–4):

■ Identifying issues
■ Identifying solutions
■ Setting goals
■ Selecting projects
■ Measuring progress

Identifying Issues

The Steering Committee is constantly involved in updating its safety and health plan. As issues emerge, they are given consideration for inclusion in the plan. One of the most effective ways of identifying critical issues is employee feedback. The amount and intensity of feedback concerning a given issue are indicators of the importance employees attach to the issue. This information, coupled with hard data, can help ensure that the Steering Committee properly prioritizes the issues it deals with when planning.

Figure 14–4
Uses of Employee Feedback

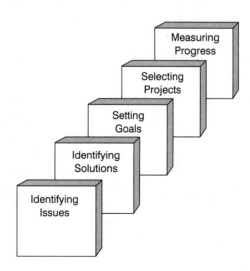

Identifying Solutions

Once the Steering Committee charters an IPT, it is looking for a solution to a problem. Consequently, it is critical that the Steering Committee listen to the feedback the IPT provides. If the IPT has done its job right, its feedback will include opinions solicited from a broad cross-section of stakeholders (e.g., employees who are affected by the problem in question). This broad-based employee feedback will, in turn, help the Steering Committee make the best decision possible when considering options for solutions.

Setting Goals

An organization's safety and health plan is a living document that is used constantly and updated continually. Have we achieved goals that appear in the plan? Do we need to add new goals, or revise existing goals? These are questions the Steering Committee should ask itself constantly. Employee feedback, if properly used, will help provide answers to these questions.

Selecting Projects

Good enough is never good enough in a TSM setting. The quality of the work environment can and should be improved continually, forever. There will always be IPTs working on projects relating to safety and health concerns. Consequently, the Steering Committee will always be concerned with selecting projects. Employee feedback is an excellent source of intelligence for the Steering Committee as it considers new projects and assigns them a priority.

Measuring Progress

Employee feedback is an essential part of the Check component of the A-P-D-C-A cycle as applied by the Steering Committee. Coupled with hard data concerning such factors as injury rates, lost time due to accidents, and medical costs, employee feedback can give the Steering Committee the information it needs to measure the progress of IPTs and make adjustments as necessary.

CAUTIONS CONCERNING EMPLOYEE FEEDBACK

There are two concerns related to employee feedback about which the Steering Committee should be well informed. The first concern is confidentiality. If anonymity is not carefully guarded when collecting employee feedback, two things will happen: (a) the feedback provided will be so bland and circumspect as to be meaningless; and (b) the feedback stream will soon run dryer than a west Texas river during a drought.

Surveys administered to employees should be designed to protect the respondent's anonymity. When the TSM Facilitator reports employee feedback to the Steering Committee, there should be no discussion of who said what. Rather, the focus should be exclusively on what was said. Any approach to collecting employee feedback that does not satisfy the confidentiality criterion is likely to be ineffective.

The second concern when collecting employee feedback is the shoot-the-messenger syndrome. When presented with bad news, some people respond by lashing out at the one delivering the news. This is the shoot-the-messenger syndrome. After all, nobody likes to hear bad news. For managers, few mistakes are more serious than shooting the messenger. When managers threaten or strike out at the bearer of bad news they run the risk of cutting themselves off from their most valuable resource—information. After all, if bringing bad news puts an employee at risk, human nature says "Don't bring it."

The example of John Pruitt, a plant manager who has a well-earned reputation for shooting the messenger, illustrates the danger inherent in making this mistake. Pruitt's plant manufactures watertight containers for the federal government. The containers are used for shipping munitions systems such as smart bombs and air-to-ground missiles.

The government specifications for these containers allow vendors to finish them with two types of paint. In order to maximize profits, Pruitt has always required his purchasing manager to order the less expensive of the two; a paint that contains several highly toxic ingredients. Because of its toxicity, the paint must be carefully handled, stored, and applied. Environmental regulations in the community where Pruitt's plant is located are tough, and they are strictly enforced.

Consequently, Amos Burch, the paintshop supervisor, knew that the plant had a problem when he found a 50-gallon drum of the toxic paint that had been accidentally punctured. Before Burch could seal off the leak, more than 30 gallons of paint had spilled. Because the paint storage bin is outside and uses wooden pallets for flooring, most of the spilled paint seeped into the ground.

Local environmental regulations require that such spills be reported and cleaned up immediately. But Burch, afraid to tell Pruitt because of his propensity for shooting the messenger, stalled. Unfortunately, before Burch worked up the courage to inform Pruitt of the spill, a county environmental compliance officer decided to pull a surprise audit. The fine levied against the company was described in the local newspaper as "substantial."

=========== TSM CASE STUDY ===

What Would You Do?

Barbara Davis is in a quandary. The Chair of the Steering Committee, her company's CEO, is upset about some of the responses to the recent employee perceptions survey. Although what several employees said is true, their feedback was blunt, to-the-point and anything but tactful. Even so, Davis is strongly opposed to tying feedback to specific names. Somehow, she must convince the Steering Committee Chair to back down from his demand for names. If you were in Davis's situation, what would you do?

============ STEP 14 IN ACTION ===

Mack Parmentier felt good about the progress MPC had made in adopting TSM. IPTs were currently working, others had completed their work and been disbanded, and yet others were in the process of being chartered. Parmentier monitored the data relating to several key indicators constantly. He also solicited employee feedback continually and put it to good use. As a result, employee perceptions of the work environment at MPC have improved steadily.

It is employee perceptions that concern Parmentier at the moment. Major ergonomic improvements have been made in the company's CAD/CAM unit as a result of the recommendations of an IPT. Parmentier is anxious to know what the CAD/CAM technicians at MPC think of the improvements.

Because the entire unit consists of just twelve employees, Parmentier has ruled out using a written survey to solicit feedback. Instead, he plans to schedule a focus group meeting and invite all twelve CAD/CAM employees to participate. Parmentier is partial to the focus-group approach and has learned to use it effectively. Parmentier has developed a survey-like document he calls his discussion guide. He will use it to get the ball rolling when the focus group meets, and, once the discussion has started, to keep it on track.

Parmentier is looking forward to the feedback he knows will be offered. From past experience he knows that MPC's employees have begun to realize that TSM is real, and that it works. Most employees have been involved in the work of at least one IPT. They have seen the results of their efforts, and they have seen that the Steering Committee is serious about making MPC's work environment safe, healthy, and productive.

SUMMARY

1. The feedback loop consists of the various communication vehicles that connect the Check component of the A-P-D-C-A model with the Plan component.

2. Continual feedback from IPTs to the Steering Committee is important because: (a) the Steering Committee is ultimately responsible for safety and health in the organization; (b) new teams are unknown, inexperienced entities; (c) the Steering Committee must ensure that the charter is being carried out.

3. The frequency of feedback from IPTs to the Steering Committee is a function of two factors: (a) experience of the IPT members, and (b) gravity of the problem the IPT is dealing with.

4. Prompt feedback from the Steering Committee to IPTs is critical. Often an IPT will be unable to move forward until it receives feedback from the Steering Committee. The morale of team members may begin to suffer if feedback is slow to come and given only sporadically.

5. In a TSM setting, a good rule of thumb concerning employee attitudes is, "Don't assume, measure." To find out if improvements are making a difference in employee attitudes, solicit feedback from them. Employee feedback collected periodically will let the Steering Committee know if its efforts are having the desired effect.

6. The most widely used methods for collecting feedback concerning employee perceptions are as follows: survey (internal), focus groups (internal), survey (external), and focus groups (external).

7. Employee feedback should be used when identifying issues, identifying solutions, setting goals, selecting projects, and measuring progress.

8. Two cautions are important when soliciting employee feedback: (a) confidentiality must be guaranteed; and (b) managers must avoid the shoot-the-messenger syndrome.

KEY TERMS AND CONCEPTS

A-P-D-C-A model	Identifying solutions
Acknowledgment	Measuring progress
Disapproval	Partial approval
Feedback loop	Selecting projects
Identifying issues	Setting goals

REVIEW QUESTIONS

1. Define the term *feedback loop*.
2. What is the rationale for IPT feedback to the Steering Committee?
3. What factors determine how frequently IPTs should provide feedback to the Steering Committee?
4. What is the rationale for Steering Committee feedback to IPTs?
5. Why is employee feedback so important in a TSM setting?
6. List and describe the methods used most frequently for collecting feedback concerning employee perceptions.
7. Explain the cautions that should be observed when soliciting employee feedback.

ENDNOTES

1. David L. Goetsch, *Occupational Safety and Health in the Age of High Technology*, 2nd ed., (Upper Saddle River, N.J.: Prentice Hall, 1996), p. 513.
2. J. Main, "Here Comes the Big Clean-Up," *Fortune*, Vol. 118, No. 12 (November 21, 1988), p. 102.

Establish a TSM Culture

- TSM Culture Defined
- Recognizing a TSM Culture
- Identifying and Removing Organizational Roadblocks
- Step 15 in Action

The previous steps have been devoted to implementing the TSM concept. The Steering Committee is in place and functioning, the TSM Facilitator is actively engaged, and IPTs are at work throughout the organization. As a result, the work environment is being improved continually.

The purpose of this step, the final step, is to make sure that the TSM approach becomes the norm. In other words, the purpose of Step 15 is to establish a TSM Culture.

TSM CULTURE DEFINED

Even with the Steering Committee, TSM Facilitator, and IPTs in place and working, the future of TSM may still be in doubt. The fourteen fundamental implementation steps have been completed, and this is a major accomplishment. But there still may be barriers, mostly hidden, to the long-term success of TSM. This is because TSM, as an approach to ensuring a safe, healthy, and productive work environment, is still in its infancy. It hasn't yet matured and become the normal and natural way of doing things. In other words, TSM is not yet part of the organizational culture.

In order to understand the meaning of TSM culture, one must first understand the concept of organizational culture. Every organization has one. An organization's culture is the everyday manifestation of its most strongly held values and beliefs. It can be seen in how employees at all levels behave on the job, their expectations of the organization and of each other, and what is considered normal and acceptable in terms of how employees approach their jobs. What an organization truly values will show up in the

215

everyday behavior and performance of employees at all levels, and no amount of lip service and sloganeering to the contrary will change this.

For example, have you ever been treated poorly or received bad service in an establishment that displays such slogans as "The customer is always right," or "Customer satisfaction is guaranteed." It is an all-too-common experience. Such establishments talk the talk, but don't walk the walk, so to speak. Customer satisfaction is a part of their marketing campaign, but not part of their culture.

With this background, the concept of organizational culture can now be defined as it relates specifically to TSM. The definition is as follows:

> *A TSM culture is the everyday manifestation of a deeply ingrained set of values that makes continually improving the work environment one of the organization's highest priorities. It shows up in procedures, expectations, habits, and traditions that promote safety, health, and competitiveness.*

RECOGNIZING A TSM CULTURE

The concept of the TSM culture has been defined, but how does one know when such a culture is in place? A frequently asked question is as follows: "How will I know a TSM culture when I see one?" This question is answered best by applying the old adage, "It's not what you say that counts, it's what you do." Figure 15–1 is a checklist of characteristics shared by organizations that have a TSM culture. If an organization has an ingrained, mature TSM culture, it will have the characteristics shown in this checklist. It is a simple task to determine if an organization's behavior matches its slogans. The following story illustrates the concept of a TSM culture.

A Tale of Two Companies

Two companies decided to implement TSM as a way to enhance their competitive standing in the marketplace. Both have now gone through all fifteen steps in the implementation process, and sufficient time has elapsed for TSM to have taken hold. In Company A, TSM is now part of the organizational culture. It is the normal way of doing business. However, in Company B, TSM is sputtering along with mixed results that are sporadic at best.

Company A assigns the development of safety slogans to IPTs. New slogans are adopted and displayed monthly. All employees at all levels model the behavior advocated in the slogans. Each month when the safety slogan is changed, executive managers make a point of visibly setting an example of not just talking the talk, but walking the walk. In fact, the CEO once sent a memo to all employees informing them that he had assessed himself a $50 fine to be contributed to the employees' social kitty for entering the production facility without first donning a hard hat.

It had been done after hours, nobody else was present to observe the infraction, and he had simply gotten in a hurry and forgotten. Nevertheless, when the CEO realized what he had done, he decided to assess himself the maximum fine. You see, the safety slogan that particular month had been Zero Tolerance of Safety Infractions.

Figure 15–1
TSM Cultural Characteristics
Checklist

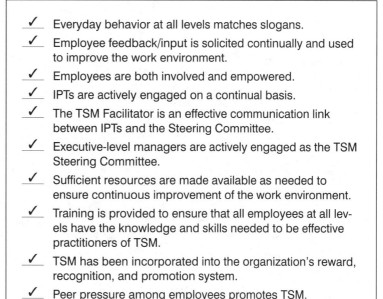

✓ Everyday behavior at all levels matches slogans.

✓ Employee feedback/input is solicited continually and used to improve the work environment.

✓ Employees are both involved and empowered.

✓ IPTs are actively engaged on a continual basis.

✓ The TSM Facilitator is an effective communication link between IPTs and the Steering Committee.

✓ Executive-level managers are actively engaged as the TSM Steering Committee.

✓ Sufficient resources are made available as needed to ensure continuous improvement of the work environment.

✓ Training is provided to ensure that all employees at all levels have the knowledge and skills needed to be effective practitioners of TSM.

✓ TSM has been incorporated into the organization's reward, recognition, and promotion system.

✓ Peer pressure among employees promotes TSM.

✓ Organizational heroes are effective practitioners of TSM.

✓ TSM is viewed by employees as the natural, normal, and acceptable way of doing things.

In Company B, safety slogans are changed infrequently and observed inconsistently. Rules relating to slogans are sometimes enforced, but typically only after an accident or incident occurs, and then only until the accident-induced surge of interest in safety has run its course. Several key managers behave as if safety slogans and rules apply to everyone but them.

In Company A, IPTs are at work all the time. It would be a rare day when at least one IPT could not be found working to improve some aspect of the work environment. In Company B, IPTs are activated only in response to a specific stimulus such as an accident, an incident, or a near miss. Company A is proactively engaged in improving the work environment, while Company B is reactively engaged. As a result, Company A often anticipates and solves problems before they cause an accident. Company B simply responds to the situation after the damage has already been done.

In Company A, employee performance relative to safety and health is evaluated as part of the performance appraisal process. Participation in TSM-related activities is factored in when selecting employees for special recognition (such as employee of the month) or rewards (promotions or raises, for example). In Company B, there is no correlation between TSM-related performance and rewards and recognition. The performance appraisal process is silent on the subject of safety and health. Evaluation forms contain no criteria relating to safety, health, or TSM. TSM-related behavior is talked about, but it is not measured, recognized, or rewarded.

In Company A, TSM is the normal and expected way of doing things. Working on improvement projects as part of an IPT is considered a regular part of the job. Peer expectations tend to reinforce TSM-positive behaviors and reject those that mitigate against safety and health. In company B, TSM-positive behaviors still seem to be unnatural and forced. There do not seem to be strong peer expectations concerning TSM-positive behaviors.

Company A has a TSM culture. Company B does not. One could simply walk through each of the companies and observe the difference. The reasons behind the differences that would be observed are typically visible, and, in most cases, measurable.

IDENTIFYING AND REMOVING ORGANIZATIONAL ROADBLOCKS

Just as the IPT is a critical component in the implementation of TSM, it can also be a critical component in establishing a TSM culture. TSM represents cultural change. Any time an existing culture is changed there will be leftover organizational roadblocks that must be identified and eliminated.

=========== TSM CASE STUDY ================================

What Would You Do?

Shelley McGuire is having trouble with an important member of Kel-Tran's Steering Committee, John Blue. Blue cannot understand why several employees in the Fluid Reclamation Department are behaving as if their operating procedures had not been revised. The new procedures were implemented to reduce hazards to employees, yet some employees seem to be ignoring them. McGuire needs to help Blue understand that this is normal and to be expected. If you were in her place, what would you do?

Identifying Organizational Roadblocks

Operating procedures, reward and recognition systems, and administrative processes are the most common sources of organizational roadblocks. A roadblock in the current context is any factor that works against the successful establishment of a TSM culture. It might be a rule, regulation, procedure, habit, attitude, or a long-standing but unwritten way of doing things. The IPT can be an excellent vehicle for rooting out and eliminating organizational roadblocks.

The charter given to an IPT assigned the task of rooting out cultural roadblocks would revolve around the following question: "What existing rules, regulations, procedures, habits, or other factors are inhibiting the effective practice of TSM?" IPTs given such a charter must by necessity be cross-functional in their makeup.

Figure 15–2 is a checklist of strategies IPTs can use for identifying roadblocks. Individual IPT members can interview employees in their respective departments and/or employees from other departments. The IPT can survey employees organization-wide.

Figure 15–2
Strategies for Identifying Cultural
Roadblocks

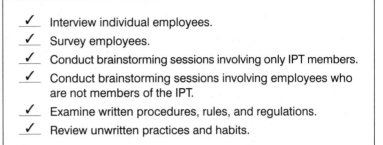

✓ Interview individual employees.

✓ Survey employees.

✓ Conduct brainstorming sessions involving only IPT members.

✓ Conduct brainstorming sessions involving employees who are not members of the IPT.

✓ Examine written procedures, rules, and regulations.

✓ Review unwritten practices and habits.

Figure 15–3 is an example of a survey instrument that might be used to identify cultural roadblocks. The same document can be used as a discussion guide when conducting brainstorming sessions aimed at identifying roadblocks.

In examining written procedures, rules, and regulations, the IPT should look for factors that encourage behaviors that are contrary to the TSM philosophy, or that although not contrary are nonsupportive. An example of a contrary factor is an outdated safety procedure describing the proper way to lift a certain type of box. The method described in this regulation, although accurate when written, does not require the use of a back support vest. One of the first recommendations of the company's first IPT was that back-support vests be worn by all employees who lift materials manually. However, the lifting-procedure section of the company's safety manual has not been updated, nor have the old habits that are based on this procedure.

Figure 15–3
TSM Cultural Roadblocks
Survey

1. Are you aware of any operating procedures that make it difficult to properly practice TSM?

2. Are you aware of any company rules or regulations that make it difficult to properly practice TSM?

3. Are you aware of any habitual behaviors (behaviors we continue only out of habit) in our company that make it difficult to practice TSM? If yes, describe them.

4. Are you aware of any unwritten rules or expectations that make it difficult to properly practice TSM? If yes, describe them.

5. Does our organizational structure (departments, divisions, etc.) support or inhibit the proper practice of TSM? If it inhibits, explain how.

6. Does our recognition system (employee-of-the-month, etc.) support or inhibit the proper practice of TSM? If it inhibits, explain how.

An example of a nonsupportive factor is a performance appraisal form that has no TSM-related criteria. Such a factor does not actively work against TSM, but does so passively because it fails to encourage TSM-supportive behavior. If employees know that TSM-supportive behavior will improve their performance rating, they will be more likely to adopt these behaviors.

Removing Organizational Roadblocks

As with all aspects of TSM, the A-P-D-C-A model is important when attempting to remove organizational roadblocks. The Assess step is accomplished when IPTs identify organizational roadblocks. The next step involves developing a plan. This is a critical step. Just because a roadblock has been identified doesn't mean it will go away. Cultural roadblocks must be carefully, intentionally, and systematically removed from an organization.

Figure 15–4 contains a set of guidelines that can be used when developing a plan for removing organizational roadblocks. The questions asked in Figure 15–4 should be answered in writing. The written answers then become the plan for removing the roadblocks in question. This may seem like a lot of trouble to go to, but remember the roadblocks in question are deeply ingrained and cultural. Such roadblocks have probably produced habitual behaviors in employees, and habits, as anyone knows, are hard to break. Trying to remove cultural roadblocks is the organizational equivalent of trying to stop smoking.

The plan developed according to the guidelines in Figure 15–4 will contain all of the information needed to root out and eliminate cultural roadblocks in an organization. However, even with the best plan it may take time to completely eliminate a roadblock.

Figure 15–4
Planning Guidelines for
Removing Cultural Roadblocks

- Who are the stakeholders? Who will be affected by the change?
- Who is likely to oppose the change?
- Who must be involved in order for the change to succeed?
- What specific tasks must be accomplished in order to remove the roadblock?
- What are the most likely barriers that might prevent the accomplishment of these tasks?
- What related processes and procedures might be affected by the removal of the roadblocks in question?
- When should the roadblock be removed?
- How will we know that the roadblock has been removed?
- What are the expected benefits of removing the roadblock in question?

When implementing the plan, it may be necessary to apply some or all of the following strategies:

- Understand the emotional transition process.
- Turn key people into advocates.
- Take a hearts-and-minds approach.
- Apply courtship strategies.

Emotional Transition Process

People confronted by cultural change often undergo a transitional process, as illustrated in Figure 15–5. The amount of time that must elapse between Steps 1 and 7 of the process will vary from person to person, but most people go through all seven steps. Consequently, if a TSM initiative is causing cultural change, the TSM Facilitator and members of the Steering Committee need to understand that acceptance on the part of employees may be delayed. Don't look for or expect immediate acceptance of change. Rather, keep working the issue in question and give the emotional transition process time to evolve.

Turn Key People into Advocates

In every situation, there are individuals who can facilitate and those who can inhibit the implementation of change. This power can come from a variety of sources (e.g., authority of position, strength of personality, respect and esteem of colleagues). These powerful employees, regardless of their official positions in the organization, must be part of the change process in order for it to work.

Identify these influential employees, bring them together, and give them the change plan. Give them an opportunity to voice concerns and raise issues. Hear their concerns, speak to their issues, and then make it clear that their cooperation is both needed and expected. A tactic that can be effective is to make these influential employees responsible for the success of the implementation. Executive managers must use their own judge-

---

TSM TIP

Enforcing Safety Rules

"Objectivity and consistency are critical when enforcing rules. Objectivity means the rules are enforced equally regardless of who commits an infraction. . . . Consistency means that rules are enforced in the same manner every time with no regard to any outside factors."[1]

Figure 15–5
Emotional Transition Process

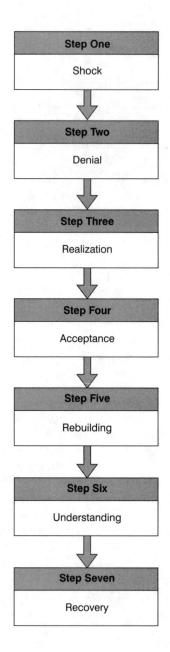

TSM TIP

Gaining a Commitment to Safety

An effective way to gain the commitment of employees is to ask them to sign a commitment document. Organizations gain three advantages from asking employees to sign on the bottom line:

1. *A signature is visual evidence of a personal commitment.*
2. *A signature is a promise to interact positively with fellow employees when they ignore safety precautions.*
3. *A signature gives other employees permission to correct the signatory's behavior when he or she ignores safety precautions.*[2]

---

ment in applying the appropriate amounts of persuasion and pressure (the carrot and the stick) with influential employees.

Take a Hearts and Minds Approach

Employees may understand the need for change on an intellectual level, but still resist it on an emotional level. Cultural change has more to do with emotion (heart) than cognition (mind). Consequently, the TSM Facilitator and members of the Steering Committee should be prepared to invest the time necessary to deal with the emotional reactions of employees to cultural change.

The most effective strategy in this regard is open face-to-face communication. Don't cover up or avoid resistance to the change in question. Rather, let resisters voice their concerns, including even outright objections, then respond to their concerns in an open, objective, nondefensive manner. Advocates of TSM have the high ground. After all we are talking about changes that are in the long-term best interests of employees and the organization. When you have seized the high ground, use it.

Once key employees accept the change, others will begin to fall in line. Eventually, the majority of employees will become proponents. When this happens, peer pressure and working with the advantage of critical mass will operate in favor of the desired change.

Apply Courtship Strategies

Courtship is an interpersonal process that moves deliberately, if sometimes slowly, toward a desired end. During courtship, the partner hoping to achieve the desired end (marriage) states his case honestly, but tactfully, and listens patiently while his partner deliberates, expresses doubts, and raises concerns. If advocates of cultural change think of their relationship with potential resisters as a courtship, they will be more successful in bringing them along and eventually winning them over.

Looking back over his first year at MPC, Mack Parmentier can reflect with pleasure on what could only be seen as a success story. Brought into a difficult situation by a reluctant CEO, he had played a key leadership role in turning things around. Not only had TSM now been fully implemented, it was rapidly becoming the cultural norm at MPC. Medical costs are down and attendance is up. Lost days due to accidents are down and morale is up. Quality problems are down and productivity is up.

Parmentier had been confident from the outset that MPC could and would implement TSM. Conditions at the company had become so bad that MPC's managers and employees really had no choice. At the outset, Parmentier had made it clear that MPC had only two options; change or go out of business.

The company had implemented TSM, but would it stick? Parmentier knows that organizations sometimes relax once they begin to experience success. For the past month Parmentier has been concerned about what he calls the pushing-the-ball-uphill syndrome. If you push a ball up a hill, the minute you relax and stop pushing, the ball will begin to roll back down the hill. Once the ball starts rolling backwards, it can be difficult to stop. Regaining uphill momentum can be difficult if not impossible. The concept of inertia works against you. Consequently, when trying to make cultural change in an organization, one must keep going until "the ball is over the summit." Once the ball begins to roll downhill on the other side, it gains momentum and inertia works in your favor.

Parmentier wants to make sure that TSM gets over the summit of the cultural hill at MPC. He knows that TSM is near the top of the mountain, but it is not yet over the top. To give TSM that critical final push, Parmentier must root out and eliminate all remaining cultural inhibitors. Working with the Steering Committee and an IPT that includes representatives from all elements of the workforce—line management, executive management, and the human resources department—Parmentier has helped revise MPC's reward and recognition system. The system is now supportive of TSM. He is currently coordinating the work of several IPTs that are attempting to identify administrative processes and operating procedures that promote behaviors contrary to TSM.

Parmentier is comfortable that most of the TSM-inhibitive processes and procedures at MPC have been identified. Plans to make the necessary on-paper changes are already being drafted. What has Parmentier concerned are residual behaviors growing out of long-standing processes and procedures, behaviors that are habitual and ingrained. He knows that changing processes and procedures on paper, although this step is important, does not always result in corresponding changes in employee behavior.

It is this point that Parmentier is preparing to make to the Steering Committee. Because of the success the company has experienced in implementing TSM, Steering Committee members seem to be relaxing somewhat. Because plans are currently being developed to eliminate process-and-procedure inhibitors, Steering Committee members seem to think that the battle is won. Parmentier's challenge is two-fold.

First, he needs to help the Steering Committee understand the pushing-the-ball-uphill syndrome. Second, he needs to help the committee understand the emotional transition process employees go through when asked to change. Parmentier knows that if all Steering Committee members understand these two concepts, they will do what is necessary to take the final step that will make TSM the cultural norm at MPC.

SUMMARY

1. An organizational culture is the everyday manifestation of its most strongly held values and beliefs.

2. A TSM culture is the everyday manifestation of a deeply ingrained set of values that make continually improving the work environment one of the organization's highest priorities. It shows up in procedures, expectations, habits, and traditions that promote safety, health, and competitiveness.

3. The TSM culture in an organization can be seen not in its slogans, but in its behavior, expectations, habits, and traditions.

4. The IPT is an excellent tool for identifying cultural impediments to TSM. These impediments are typically found in an organization's administrative processes and operating procedures.

5. Strategies for removing cultural impediments to TSM include the following: understand the emotional transition process, turn key people into advocates, take a hearts and minds approach, and apply courtship strategies.

KEY TERMS AND CONCEPTS

Acceptance   Realization

Courtship strategies   Rebuilding

Denial   Recovery

Emotional transition process   Shock

Hearts-and-minds approach   TSM culture

Organizational culture   Understanding

REVIEW QUESTIONS

1. Define the term *organizational culture*.

2. Define the term *TSM culture*.

3. Explain the key to recognizing a TSM culture.

4. Where are the best places to look when attempting to identify cultural impediments to TSM?

5. List and explain four strategies for removing organizational roadblocks to TSM.

ENDNOTES

1. David L. Goetsch, *Occupational Safety and Health in the Age of High Technology*, 2nd ed. (Upper Saddle River, N.J.,: Prentice Hall, 1996), p. 405.

2. A. Fettig, "Sign Up for Safety," *Safety and Health*, Vol. 144, No. 1 (July 1991), p. 26.

Hazard-Prevention Checklists

Mechanical Hazards

- Are point-of-operation guards being properly used where appropriate?
- Are point-of-operation devices being properly used where appropriate?
- Are the safest feeding/ejection methods being employed?
- Are lockout/tagout systems being employed as appropriate?
- Are automatic shutdown systems in place, clearly visible, and accessible?
- Are safeguards properly maintained, adjusted, and calibrated as appropriate?

Fall-Related Hazards

- Are foreign objects present on the walking surface or in walking paths?
- Are there design flaws in the walking surface?
- Are there slippery areas on the walking surface?
- Are there raised or lowered sections of the walking surface that might trip a worker?
- Is good housekeeping being practiced?
- Is the walking surface made of or covered with a nonskid material?
- Are employees wearing nonskid footwear as appropriate?
- Are ladders strong enough to support the loads they are subjected to?
- Are ladders free of cracks, loose rungs, and damaged connections?
- Are ladders free of heat damage and corrosion?
- Are ladders free of moisture that can cause them to conduct electricity?
- Are metal ladders free of burrs and sharp edges?
- Are fiberglass ladders free of blooming damage?
- Are employees wearing head and face protection as appropriate?
- Are employees wearing foot protection as appropriate?

Lifting Hazards

- Are posters that illustrate proper lifting techniques displayed strategically throughout the workplace?
- Are machines and other lifting aids available to assist employees in situations where loads to be lifted are too heavy and/or bulky?
- Are employees who are involved in lifting using personal protective devices?

Heat and Temperature Hazards

- Are workers in hot environments being gradually acclimatized?
- Are workers in hot environments rotating into cooler environments at specified intervals?
- Are workers in hot environments wearing personal protective clothing?
- Are first aid stations and supplies readily available for the treatment of burn victims?
- Are eye wash and emergency shower stations readily available for chemical burn victims?
- Are central or spot heating furnaces, warm air jets, contact warm plates, or radiant heaters being used where employees work in cold environments?
- Are work areas in cold environments shielded from the wind?
- Are the handles of tools used in cold environments covered with insulating material?
- Are metal chairs in cold environments covered with insulating material?
- Are heated tents or shelters provided so that workers in cold environments can take periodic warming breaks?
- Are appropriate types of warmed drinks made available to workers in cold environments?
- Are jobs done in cold environments designed to require movement and minimize sitting or standing still?
- Are jobs done in cold environments designed so that as many tasks as possible are performed in a warm environment?

Pressure Hazards

- Are boilers properly installed (level, sufficient room around them for inspecting all sides, etc.)?
- Are control/safety devices on boilers present and in proper working condition?
- Is a schedule of regular inspections for all boilers posted and clearly visible?
- Is a schedule of preventive maintenance for all boilers posted and adhered to?

- Are pulse dampening devices/strategies being employed on high pressure systems?
- Are high pressure systems installed with a minimum of joints?
- Are appropriate pressure gauges in place and working properly?
- Are shields placed around high pressure systems?
- Are remote control and monitoring devices being used with high pressure systems?
- Is access restricted in areas where high pressure systems are present?
- Are leak detection methods being employed with pressurized gas systems?

Electrical Hazards

- Are short circuits present anywhere in the facility (use a professional electrician to run the necessary checks)?
- Are static electricity hazards present anywhere in the facility?
- Are electrical conductors in close enough proximity to cause an arc?
- Are explosive/combustible materials stored or used in proximity to electrical conductors?
- Does the facility have adequate lightning protection?
- Has insulation degradation resulted in bare wires anywhere in the facility?
- Is all electrical equipment in proper working order (use a professional electrician to run the necessary checks)?

Fire Hazards

- Are automatic fire detection systems employed in the facility?
- Are flammable materials and substances stored in a way that isolates them from oxygen and heat?
- Are automatic fire extinguishing systems employed in the facility? Are fixed extinguishing systems strategically located throughout the facility?

Toxic Substance Hazards

- Are all hazardous substances clearly identified and marked?
- Are all toxic substances properly stored?
- Are persons handling toxic substances using appropriate personal protective equipment that is properly maintained?
- Is access to areas that contain toxic substances appropriately restricted?
- Is equipment used to produce, process, and/or package toxic substances properly maintained?

Explosive Hazards

■ Is smoking prohibited in the vicinity of explosive materials?

■ Are methods employed to eliminate static electricity?

■ Are spark-resistant tools employed wherever possible?

■ Are processes that use explosive substances segregated in stand-alone facilities?

■ Are processes and facilities that use explosive materials well ventilated?

■ Are automatic shut-down systems employed on all processes that use explosives?

■ Are all areas where explosives are present free of ignition sources?

■ Is good housekeeping being practiced in all areas where explosives are present?

Radiation Hazards

■ Are personal monitoring devices being used where appropriate?

■ Are radiation areas clearly marked with caution signs?

■ Are high radiation areas clearly marked with caution signs?

■ Are airborne radiation areas clearly marked with caution signs?

■ Are all containers in which radioactive materials are stored and/or transported clearly marked with caution signs?

■ Is an evacuation warning signal system in place and operational?

■ Are all containers in which radioactive materials are stored secured against unauthorized removal?

Vibration Hazards

■ What are the sound level meter readings taken at different times and locations?

■ What are the dosimeter readings for each shift of work?

■ Are audiometer tests being conducted at appropriate intervals?

■ What are the results of audiometric tests?

■ Are records of audiometric tests being properly maintained?

■ What is being done to reduce noise at the source?

■ What is being done to reduce noise along its path?

■ Are workers wearing personal protective devices where appropriate?

■ Are low-vibration tools being used wherever possible?

■ Are employees who use vibrating tools or equipment doing the following as appropriate:

  • Wearing properly fitting thick gloves?

  • Taking periodic breaks?

- Using a loose grip on vibrating tools?
- Keeping warm?
- Using vibration-absorbing floor mats and seat covers as appropriate?

Automation Hazards

■ Are workers who use VDTs as their primary work tool employing methods to prevent eye strain?

■ Are all robot sites well lighted and marked off?

■ Are the floors around robot sites clean and properly maintained so that workers won't stop and fall into the robot's work envelope?

■ Are all electrical and pneumatic components of all robots equipped with guards and covers?

■ Are the work envelopes of all robots equipped with guards and covers?

■ Are automatic shut-off systems in place for all robots and other automated systems?

■ Are lockout systems used before workers enter a robot's work envelope?

■ Are safety fences erected around all automated systems?

■ Are the controllers for all automated systems located outside of work envelopes?

Ergonomic Hazards

■ Are tasks being performed that involve unnatural or hazardous movements?

■ Are tasks being performed that involve frequent manual lifting?

■ Are tasks being performed that involve excessive wasted motion?

■ Are tasks being performed that involve unnatural or uncomfortable postures?

■ Are tasks being performed that should be automated?

Index